Practice in CHEMISTRY

Progressive questions for AS and A level

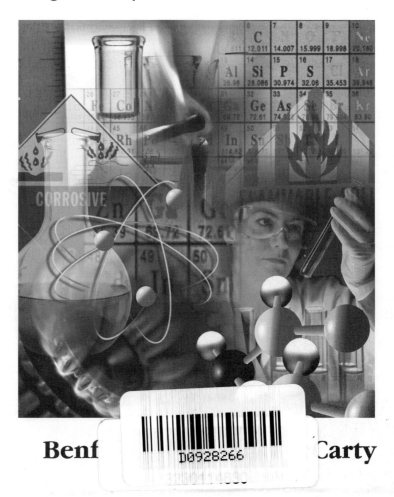

Benf D0928266 **Carty**

Hodder & Stoughton

A MEMBER OF THE HODDER HEADLINE GROUP

£9~99

Orders: please contact Bookpoint Ltd, 78 Milton Park, Abingdon, Oxon OX14 4TD.
Telephone: (44) 01235 827720, Fax: (44) 01235 400454. Lines are open from 9.00–6.00, Monday to
Saturday, with a 24 hour message answering service. Email address: orders@bookpoint.co.uk

British Library Cataloguing in Publication Data
A catalogue record for this title is available from The British Library

ISBN 0 340 77265 4

First published 2000
Impression number   10 9 8 7 6 5 4 3 2 1
Year 2006 2005 2004 2003 2002 2001 2000

NF039261
200601
(540.76)

Copyright © 2000 Frank Benfield, Robin Hillman, Colin McCarty

Cover design by Blue Pig Design Co.
Typeset by Wearset, Boldon, Tyne and Wear.
Printed in Great Britain for Hodder & Stoughton Educational, a division of Hodder Headline Plc,
338 Euston Road, London NW1 3BH by J W Arrowsmith, Bristol.

Contents

Preface

The knowledge, understanding and experimental skills required to embrace the contents of the core of the combined AS and A level Chemistry, as defined in the Subject Criteria for Chemistry, is intended to comprise about 60% of an Advanced Level Specification. This core material was the starting point for the authors of *Practice in Chemistry*. However, this constrained both the content of questions unacceptably (particularly in inorganic and organic chemistry) and what the authors believe to be the sound educational practice of extending both the breadth and the depth of knowledge appropriate to GCE candidates worthy of higher grades. A proportion of the questions therefore go beyond the limits of the core material and these are indicated by the symbol * beside the number of each question. Moreover, each of the examination groups has chosen to include different extension material beyond the basic core, a fact which should be borne in mind when teachers select questions for their students. A grid has been included to link the main content areas of each chapter with the appropriate sections of each Advanced Level specification.

A student's understanding of some topics is improved by having available a considerable number of questions of graduated difficulty, notably in quantitative chemistry. In other subject areas, reinforcement is facilitated by repetition of subject content, although the context of the question may well have been subtly changed to retain the student's interest.

The inclusion of necessarily concise answers is seen as being of assistance not only to teachers but also as a valuable tool for students. It will provide them with an opportunity to check quickly on the contents of their answers and to learn what is likely to constitute an answer worthy of high marks. Another useful revision exercise for students would be to work through a series of questions with the given answers side by side. Synoptic assessment is mandatory for Advanced GCE and is intended to test each candidate's ability to link material from more than one module. The emphasis is not on recall, but on application, analysis and synthesis: the synoptic questions in *Practice in Chemistry* have been designed to test these skills fully.

Frank Benfield, Robin Hillman, Colin McCarty

Specification matching grid

AS topics

Chapter	Topic	AQA	Edexcel	Edexcel (Nuffield)	OCR Spec. A	OCR Spec. B (Salters)
1	Amounts of substances	Module 1 (10.2.1) (10.2.2)	Unit 1 (1.2)	Unit 1 (1.4) (2.3)	Module 2811 (5.1.1)	Module 1 (EL) (DF) Module 2 (M)
1	Formulae	Module 1 (10.2.4)	Unit 1 (1.2) Unit 2 (2.2)	Unit 1 (1.3)	Module 2811 (5.1.1)	Module 1 (EL)
1	Equations	Module 1 (10.2.5)	Unit 1 (1.2)	Unit 1 (1.3)	Module 2811 (5.1.1)	Module 1 (EL) Module 2 (M)
1	Titration calculations	Module 1 (10.2.5)	Unit 1 (1.2)	Unit 1 (4.2)	Module 2811 (5.1.1)	Module 2 (M)
2	States of matter	Module 1 (10.3.4)	Unit 1 (1.3)		Module 2811 (5.1.3) Module 2815 (5.9.1)	
2	Changes of state	Module 1 (10.3.4)	Unit 1 (1.3)		Module 2811 (5.1.3) Module 2815 (5.9.1)	
2	Molecular/ giant structures	Module 1 (10.3.4)	Unit 1 (1.3)	Unit 1 (3.5)	Module 2811 (5.1.3)	Module 2 (M)
3	Ionic bonding	Module 1 (10.3.1)	Unit 1 (1.3)	Unit 1 (3.5)	Module 2811 (5.1.3)	Module 1 (EL) Module 2 (M)
4	Enthalpy changes	Module 2 (11.1.1)	Unit 2 (2.1)	Unit 1 (5.1) Unit 2 (7.6)	Module 2813 (5.3.1)	Module 1 (DF)
4	Hess's Law	Module 2 (11.1.3)	Unit 2 (2.1)	Unit 1 (5.2) Unit 2 (7.6)	Module 2813 (5.3.1)	Module1 (DF)
4	Bond energies	Module 2 (11.1.4)	Unit 2 (2.1)	Unit 2 (7.6)	Module 2813 (5.3.1)	Module 1 (DF)
5	Rates of reaction	Module 2 (11.2.1) (11.2.3)	Unit 2 (2.3)	Unit 2 (7.7)	Module 2813 (5.3.2)	Module 2 (A)
5	Catalysis	Module 2 (11.2)	Unit 2 (2.3)	Unit 2 (7.7) Unit 4 (11.5)	Module 2813 (5.3.2)	Module 1 (DF) Module 2 (A)
6	Equilibrium characteristics	Module 2 (11.3.1)	Unit 2 (2.4)	Unit 2 (7.7)	Module 2813 (5.3.3)	Module 2 (A)
6	Qualitative effects of changes of conditions on the position of equilibrium	Module 2 (11.3.2) (11.3.3)	Unit 2 (2.4)	Unit 2 (7.7)	Module 2813 (5.3.3)	Module 2 (A) Module 5 (AA)
7	Atomic structure	Module 1 (10.1.1) (10.1.2) (10.1.3)	Unit 1 (1.1)	Unit 1 (3.1)	Module 2811 (5.1.2)	Module 1 (EL)
7	Mass spectrometry	Module 1 (10.1.3)	Unit 1 (1.1)	Unit 1 (3.3) Unit 6 (21.1)	Module 2811 (5.1.1)	Module 1 (EL) Module 2 (WM)
8	Electronic configuration of atoms	Module 1 (10.1.4)	Unit 1 (1.1)	Unit 1 (3.3) (3.4)	Module 2811 (5.1.3)	Module 1 (EL) Module 2 (M)
9	Electronegativity	Module 1 (10.3.2)	Unit 1 (1.3)	Unit 2 (7.4)	Module 2811 (5.1.3)	Module 2 (A) (PR)
9	Intermolecular forces	Module 1 (10.3.3)	Unit 1 (1.3)	Unit 2 (9.1) (9.2) (9.3)	Module 2811 (5.1.3)	Module 2 (PR)
10	Shapes of molecules/ions	Module 1 (10.3.5)	Unit 1 (1.3)	Unit 2 (7.2)	Module 2811 (5.1.3)	Module 1 (EL)

AS topics

Chapter	Topic	AQA	Edexcel	Edexcel (Nuffield)	OCR Spec. A	OCR Spec. B (Salters)
11	Oxidation states	Module 2 (11.4.1) (11.4.2) (11.4.3) Module 5 (14.3.1)	Unit 1 (1.5) Unit 5 (5.1)	Unit 2 (6.2)	Module 2811 (5.1.5) Module 2815 (5.5.3)	Module 2 (M) Module 4 (SS)
12	Atomic structure and the Periodic Table	Module 1 (10.4.1)	Unit 1 (1.4)	Unit 1 (3.4)	Module 2811 (5.1.4)	Module 2 (M)
13	Group 2	Module 1 (10.4.3)	Unit 1 (1.6)	Unit 1 (4.3) (4.4)	Module 2811 (5.1.5)	Module 1 (EL)
13	Group 7	Module 2 (11.5.1) (11.5.2) (11.5.3)	Unit 1 (1.7)	Unit 2 (6.1) (6.3) (6.4) (6.5) (6.8)	Module 2811 (5.1.6)	Module 2 (M)
14	Trends in physical properties across a period	Module 1 (10.1.4) (10.4.2)	Unit 1 (1.4)	Unit 1 (3.1) (3.4)	Module 2811 (5.1.4)	Unit 1 (EL)
15	Nomenclature – alkanes and alkenes	Module 3 (12.1.1)	Unit 2 (2.2)	Unit 1 (2.1) Unit 8 (8.1)	Module 2812 (5.2.1)	Module 1 (DF)
16	Nomenclature of compounds containing a functional group	Module 3 (12.1.1) Module 4 (13.4.1)	Unit 2 (2.2)	Unit 1 (2.1)	Module 2812 (5.2.1)	Module 1 (DF) Module 2 (WM)
17	Reactions of alkanes	Module 3 (12.2.2) (12.2.3) (12.2.4)	Unit 2 (2.2)	Unit 2 (8.2)	Module 2812 (5.2.2)	Module 1 (DF) Module 2 (A)
17	Reactions of alkenes	Module 3 (12.3.1) (12.3.2)	Unit 2 (2.2)	Unit 2 (8.4)	Module 2812 (5.2.4)	Module 2 (PR)
18	Reactions of alcohols	Module 3 (12.5.1) (12.5.2) (12.5.3)	Unit 2 (2.2)	Unit 1 (2.2)	Module 2812 (5.2.5)	Module 1 (DF) Module 2 (A) (WM)
19	Reactions of halogenoalkenes	Module 3 (12.4.1) (12.4.2)	Unit 2 (2.2)	Unit 2 (10.1) (10.2)	Module 2812 (5.2.6)	Module 2 (A)

A2 topics

Chapter	Topic	AQA	Edexcel	Edexcel (Nuffield)	OCR Spec. A	OCR Spec. B (Salters)
20	Lattice enthalpies	Module 5 (14.1.1)	Unit 4 (4.1)	Unit 6 (16.2)	Module 2815 (5.5.1)	Module 5 (O)
20	Born-Haber cycles	Module 5 (14.1.1)	Unit 4 (4.1)	Unit 6 (16.1) (16.2) (16.3)	Module 2815 (5.5.1)	Module 5 (O)
21	Measurement of rate		Unit 5 (5.4)	Unit 4 (11.1)		Module 4 (EP)
21	Order of reaction	Module 4 (13.1.1) (13.1.2)	Unit 5 (5.4)	Unit 4 (11.2)	Module 2816 (5.11.1)	Module 4 (EP)
21	Mechanism and rate		Unit 5 (5.4)	Unit 4 (11.3)	Module 2816 (5.11.1)	
22	The Equilibrium Law and its applications	Module 4 (13.2.1)	Unit 4 (4.3)	Unit 4 (14.1) (14.2)	Module 2816 (5.11.2)	Module 4 (EP)
22	Applications extended to reactions involving gases	Module 4 (13.2.1)	Unit 4 (4.3)	Unit 4 (14.3)	Module 2816 (5.11.2)	Module 5 (AA)
22	Calculation of equilibrium constants	Module 4 (13.2.1) (13.2.2)	Unit 4 (4.3)	Unit 4 (14.2) (14.3)	Module 2816 (5.11.2)	Module 4 (EP) Module 5 (AA)
22	Applications of Le Chatelier's Principle	Module 4 (13.2.2)	Unit 4 (4.3)	Unit 4 (14.4)	Module 2816 (5.1.2)	Module 4 (EP) Module 5 (AA)
22	Strong and weak acids	Module 4 (13.3.4)	Unit 4 (4.4)	Unit 4 (14.6)	Module 2813 (5.3.3) Module 2816 (5.11.3)	Module 5 (O)

A2 topics

Chapter	Topic	AQA	Edexcel	Edexcel (Nuffield)	OCR Spec. A	OCR Spec. B (Salters)
22	pH calculations	Module 4 (13.3.2)	Unit 4 (4.4)	Unit 4 (14.5) (14.6)	Module 2816 (5.11.3)	Module 5 (O)
22	K_a	Module 4 (13.3.5)	Unit 4 (4.4)	Unit 4 (14.6)	Module 2816 (5.11.3)	Module 5 (O)
22	Buffer solutions	Module 4 (13.3.7)	Unit 4 (4.4)	Unit 4 (14.8)	Module 2816 (5.11.3)	Module 5 (O)
23	Periodic trends in elements, chlorides and oxides	Module 5 (14.2.1) (14.2.2) (14.2.3)	Unit 4 (4.2)	Unit 6 (16.4)	Module 2815 (5.5.2)	Module 5 (AA)
24	Transition elements	Module 5 (14.3.1) (14.4.1) (14.4.2) (14.4.3) (14.4.4) (14.4.5) (14.4.6) (14.4.7)	Unit 5 (5.2)	Unit 6 (19.1) (19.2) (19.3) (19.4)	Module 2815 (5.5.3) (5.10.2) (5.10.3) (5.10.4)	Module 4 (SS)
25	Shapes of organic molecules	Module 4 (13.4.2)	Unit 4 (4.5)	Unit 2 (8.4) Unit 6 (18.2)	Module 2811 (5.1.3) Module 2814 (5.4.5)	Module 2 (PR) Module 4 (EP)
26	Arenes	Module 4 (13.6.1) (13.6.2) (13.6.3) (13.6.4) (13.6.5)	Unit 5 (5.3)	Unit 4 (12.1) (12.2)	Module 2814 (5.4.1)	Module 5 (CD)
27	Aldehydes and ketones	Module 4 (13.5.1)	Unit 4 (4.5)	Unit 4 (15.1)	Module 2814 (5.4.2)	Module 5 (MD)
28	Carboxylic acids	Module 4 (13.5.2)	Unit 4 (4.5)	Unit 4 (15.2)	Module 2814 (5.4.2)	Module 4 (DP)
29	Amines	Module 4 (13.5.3) (13.7.1) (13.7.2) (13.7.3)	Unit 4 (4.5)	Unit 6 (18.1)	Module 2814 (5.4.4)	Module 4 (DP)
29	Amides		Unit 4 (4.5)			Module 4 (DP)
30	Synthetic pathways	Module 4 (13.10)	Unit 5 (5.5)	Unit 6 (20.1)	Module 2814 (5.4.5)	Module 5 (MD)
31	Polymerization – addition	Module 4 (13.9.1)	Unit 2 (2.2) Unit 5 (5.5)	Unit 6 (8.4) (8.5) (20.2)	Module 2812 (5.2.4) Module 2814 (5.4.6)	Module 2 (PR) Module 4 (DP)
32	Polymerization – condensation	Module 4 (13.9.2)	Unit 5 (5.5)	Unit 6 (18.5)	Module 2814 (5.4.6)	Module 4 (DP)
32	Testing for functional groups	Module 4 (13.10)	Unit 5 (5.5)	Unit 6 (20.3)	Module 2812 (5.2.4) (5.2.5) (5.2.6) Module 2814 (5.4.2) (5.4.3)	Module 5 (MD)
33	Ultra-violet and visible spectroscopy		Unit 5 (5.5)	Unit 6 (21.5)	Module 2815 (5.8.4)	Module 5 (CD)
33	Infra-red spectroscopy	Module 4 (13.11.3)	Unit 5 (5.5)	Unit 4 (15.2) Unit 6 (20.3) (21.2)	Module 2814 (5.4.7) Module 2815 (5.8.5)	Module 2 (WM) Module 5 (MD)
33	Nuclear magnetic resonance spectroscopy	Module 4 (13.11.4)	Unit 5 (5.5`)	Unit 6 (20.3) (21.3)	Module 2814 (5.4.7) Module 2815 (5.8.5)	Module 4 (EP) Module 5 (MD)
33	Mass spectrometry	Module 4 (13.11.2)	Unit 5 (5.5)	Unit 6 (20.3) (21.1)	Module 2814 (5.4.7) Module 2815 (5.8.2) (5.8.5)	Module 2 (WM) Module 5 (MD)

Part 1 AS Questions

1 Formulae, equations and amounts of substances

In this section you will need to understand:

- what is meant by a mole of a substance
- what is meant by the Avogadro constant
- what is meant by relative atomic mass
- what is meant by relative formula mass
- what is meant by molar mass (mass of one mole)
- the significance of the chemical formula for a substance
- how to convert mass in grams to number of moles and vice versa
- the use of the molar gas volume
- how to use ionic charges to determine the formula of an ionic substance
- how to calculate the empirical formula of a compound from combining masses
- how to change a word equation for a reaction into a fully balanced chemical equation
- how to write fully balanced chemical equations for reactions given only limited information
- how to write fully balanced ionic equations
- how to use balanced chemical equations to calculate masses of reactants/products including examples involving percentage yields
- what is meant by a molar solution
- how to calculate the number of moles of a substance in solution
- how to convert moles of substance in solution into a mass in grams and vice versa
- how titrations are carried out and how the desired information can be calculated from the results of such experiments.

Data: In most of the questions which follow, you will need to refer to a table of relative atomic masses.

1.1 In what units do chemists measure 'amount of substance'? [1]

1.2 What is the definition of 'a mole'? [3]

1.3 The Avogadro constant has a value of 6.022045×10^{23} mol^{-1}. Explain what is meant by this statement. [2]

1.4 What is meant by:
(a) the relative atomic mass of an element [3]

(b) the relative formula mass of a substance   [3]

(c) the molar mass of a substance?   [1]

In each case, state in what units, if any, the value is given.   [2]

1.5   What is the relative formula mass of:

(a) a neon atom, Ne   [1]

(b) a dihydrogen molecule, H_2 [1]

(c) a sulphur atom, S   [1]

(d) a sulphur molecule, S_8 [1]

(e) a hydrogen iodide molecule, HI   [1]

(f) a cyanogen molecule, C_2N_2 [1]

(g) a cyclononane molecule, C_9H_{18} [1]

(h) a trimethylamine molecule, $(CH_3)_3N$ [1]

(i) lithium iodide, LiI   [1]

(j) strontium bromide, $SrBr_2$ [1]

(k) potassium dichromate(VI), $K_2Cr_2O_7$ [2]

(l) ammonium iodate(V), NH_4IO_3 [2]

(m) iron(II) chloride-4-water, $FeCl_2.4H_2O$ [2]

(n) potassium chromium(III) sulphate, $KCr(SO_4)_2.12H_2O$ [2]

(o) a quinine molecule, $C_{20}H_{24}N_2O_2$?   [2]

1.6   Find the mass of one mole of:

(a) oxygen atoms, O   [1]

(b) dioxygen molecules, O_2 [1]

(c) sulphur dioxide, SO_2 [1]

(d) strontium hydroxide, $Sr(OH)_2$ [1]

(e) dodecane, $C_{12}H_{26}$ [1]

(f) ethanoic acid, CH_3CO_2H [1]

(g) aluminium sulphate, $Al_2(SO_4)_3$ [2]

(h) sodium carbonate-10-water, $Na_2CO_3.10H_2O$ [2]

(i) copper(II) chlorate(VII)-6-water, $Cu(ClO_4)_2.6H_2O$ [2]

(j) etioporphyrin, $C_{32}H_{38}N_4$.   [2]

1.7   Find the mass of:

(a) 0.2 mol of nitrogen atoms, N   [1]

(b) 2.5 mol of dinitrogen molecules, N_2 [1]

(c) 0.001 mol of carbon dioxide molecules, CO_2 [1]

(d) 0.40 mol of sodium hydroxide, NaOH   [1]

(e) 1.5 mol of ethanoic acid molecules, CH_3CO_2H [1]

(f) 0.20 mol of magnesium hydroxide, $Mg(OH)_2$ [1]

(g) 4 mol of potassium chromate(VI), K_2CrO_4 [1]

(h) 0.02 mol of nitrobenzene molecules, $C_6H_5NO_2$ [1]

    (i) 0.75 mol of nickel-2-chloride-1-water, $NiCl_2.2H_2O$ **[2]**

    (j) 0.0001 mol of insulin molecules, $C_{254}H_{377}N_{65}O_{75}S_6$.   **[3]**

1.8 **(a)** Silver sulphate has the formula Ag_2SO_4. What mass of silver sulphate contains one mole of silver ions?   **[2]**

    (b) Aluminium hydroxide has the formula $Al(OH)_3$. Find the mass of aluminium hydroxide which contains 0.02 mol of hydroxide ions.   **[2]**

    (c) Sodium carbonate-10-water has the formula $Na_2CO_3.10H_2O$. Find the mass of it which contains 2 mol of water of crystallization.   **[2]**

1.9 Find the number of moles:

    (a) of chlorine atoms, Cl, in 7.1 g of chlorine, Cl   **[1]**

    (b) of chlorine atoms, Cl, in 7.1 g of chlorine, Cl_2 **[1]**

    (c) of dichlorine molecules, Cl_2, in 7.1 g of chlorine, Cl_2 **[1]**

    (d) in 2.56 g of sulphur trioxide, SO_3 **[1]**

    (e) in 7.1 g of decane, $C_{10}H_{22}$ **[1]**

    (f) in 12.4 g of silver sulphide, Ag_2S **[1]**

    (g) of hydroxide ions in 29.6 g of calcium hydroxide, $Ca(OH)_2$ **[2]**

    (h) of sodium ions in 780 g of sodium sulphide, Na_2S **[2]**

    (i) of water molecules in 5.16 g of calcium sulphate-2-water, $CaSO_4.2H_2O$ **[2]**

    (j) of water molecules in 5.16 g of calcium sulphate-0.5-water, $CaSO_4.0.5H_2O$ (give the answer to two significant figures).   **[2]**

Data: In 1.10 and 1.11 assume that one mole of molecules of any gas has a volume of 24 000 cm³ (24 dm³) under normal laboratory conditions.

1.10 Find the number of moles of molecules in the following gaseous volumes

    (a) 240 cm³ of dihydrogen, H_2 **[1]**

    (b) 240 cm³ of helium, He   **[1]**

    (c) 480 cm³ of carbon dioxide, CO_2 **[1]**

    (d) 480 dm³ of sulphur dioxide, SO_2 **[1]**

    (e) 1200 cm³ of methane, CH_4 **[1]**

    (f) 1.2 dm³ of ethane, C_2H_6 **[1]**

    (g) 1.2 cm³ of propane, C_3H_8 **[1]**

    (h) 96 cm³ of uranium(VI) fluoride, UF_6 **[1]**

    (i) 2.0 cm³ of vanadium(IV) iodide, VI_4 **[1]**

    (j) the air in a room of length 8 m, width 6 m and height 2 m.   **[3]**

1.11 What is the volume occupied by the gas in each of the following examples?

    (a) 4.0 g of argon, Ar   **[1]**

    (b) 4.0 g of helium, He   **[1]**

(c) 4.0 g of dihydrogen, H_2 [1]

(d) 3.2 g of methane, CH_4 [1]

(e) 0.00048 g of trioxygen (ozone), O_3 [1]

(f) 17.6 kg of carbon dioxide, CO_2 [1]

(g) 6.8 tonnes of ammonia, NH_3 [1]

(h) a mixture of 0.28 g of carbon monoxide, CO, and 1.76 g of carbon dioxide, CO_2 [3]

(i) the oxygen in 86.4 g of air, assuming that 20% of the air is oxygen and that the average molar mass of air is 28.8 g mol^{-1} [3]

(j) the products of complete combustion of 0.0192 g of methane, CH_4, measured under normal laboratory conditions. [3]

1.12 In this question you will probably need to make use of the following table of ionic charges.

Cations

+1	+2	+3
ammonium, NH_4^+	barium, Ba^{2+}	aluminium, Al^{3+}
copper(I), Cu^+	calcium, Ca^{2+}	chromium(III), Cr^{3+}
lithium, Li^+	copper(II), Cu^{2+}	iron(III), Fe^{3+}
potassium, K^+	iron(II), Fe^{2+}	
silver, Ag^+	lead(II), Pb^{2+}	
sodium, Na^+	magnesium, Mg^{2+}	
	strontium, Sr^{2+}	
	zinc, Zn^{2+}	

Anions

−1	−2	−3
bromide, Br^-	carbonate, CO_3^{2-}	nitride, N^{3-}
chlorate(I), ClO^-	oxide, O^{2-}	phosphate(V), PO_4^{3-}
chlorate(V), ClO_3^-	sulphate, SO_4^{2-}	
chlorate(VII), ClO_4^-	sulphide, S^{2-}	
chloride, Cl^-	sulphite, SO_3^{2-}	
fluoride, F^-		
hydrogencarbonate, HCO_3^-		
hydroxide, OH^-		
iodide, I^-		
nitrate, NO_3^-		
nitrite, NO_2^-		

Write down the formulae of the following ionic compounds:

(a)	zinc oxide	**(p)**	silver fluoride
(b)	potassium sulphide	**(q)**	chromium(III) nitrate
(c)	aluminium chloride	**(r)**	strontium nitrate
(d)	ammonium nitrite	**(s)**	aluminium sulphide
(e)	copper(I) bromide	**(t)**	lithium carbonate
(f)	lithium phosphate(V)	**(u)**	magnesium phosphate(V)
(g)	barium hydroxide	**(v)**	copper(I) oxide
(h)	sodium sulphate	**(w)**	copper(II) oxide
(i)	calcium chlorate(I)	**(x)**	iron(III) iodide
(j)	calcium chlorate(V)	**(y)**	ammonium phosphate(V)
(k)	potassium chlorate(VII)	**(z)**	magnesiuim nitride
(l)	magnesium hydrogencarbonate	**(aa)**	calcium sulphite
(m)	iron(III) sulphate	**(ab)**	magnesium sulphate
(n)	lead(II) nitrate	**(ac)**	potassium nitride
(o)	sodium iodide	**(ad)**	aluminium hydroxide.

[30]

1.13 Find the simplest formula (empirical formula) for each of the following compounds:

(a) an oxide of hydrogen in which 0.4 g of hydrogen are combined with 6.4 g of oxygen [2]

(b) a bromide of iron in which 1.12 g of iron are combined with 3.20 g of bromine [2]

(c) a sulphide of aluminium, 1.50 g of which are formed from 0.54 g of aluminium [2]

(d) an oxide of phosphorus, 0.89 g of which are formed from 0.39 g of phosphorus [2]

(e) a compound consisting of 4.8 g of carbon, 6.4 g of oxygen and 12.8 g of sulphur [2]

(f) a compound containing 0.90 g of carbon combined with 0.10 g of hydrogen [2]

(g) a compound containing 0.0040 g of calcium combined with 0.0028 g of silicon and 0.0048 g of oxygen [2]

(h) a compound containing 40% by mass of calcium together with 12% of carbon and 48% of oxygen [2]

(i) a compound containing 40% by mass of carbon combined with 6.7% of hydrogen and 53.3% of oxygen [2]

(j) a hydrated salt which contains 1.96 g of iron, 1.12 g of sulphur, 2.24 g of oxygen combined with 4.41 g of water of crystallization. [2]

1.14 Find the molecular formula for each of the following compounds:

(a) a compound in which 1.08 g of carbon are combined with 0.09 g of hydrogen and its relative formula mass is 78 [3]

(b) a compound in which 6.60 g of carbon are combined with 1.10 g of hydrogen and its relative formula mass is 70 **[3]**

(c) a compound in which 1.68 g of phosphorus reacts with oxygen to form 2.98 g of product and the molar mass of the substance is 220 g mol^{-1} **[3]**

(d) a substance which has a molar mass of 60 g mol^{-1}. On analysis a sample of it was found to contain 0.84 g of carbon, 1.12 g of oxygen, 1.96 g of nitrogen and 0.28 g of hydrogen **[3]**

(e) 0.02 moles of a substance has a mass of 0.94 g and this mass contains 0.80 g of mercury, the rest being chlorine. **[3]**

1.15 Turn each of the following word equations into fully balanced chemical equations using appropriate symbols and formulae (do not include state symbols):

(a) magnesium + oxygen → magnesium oxide **[3]**

(b) calcium carbonate → calcium oxide + carbon dioxide **[3]**

(c) hydrogen + oxygen → water **[3]**

(d) magnesium + hydrochloric acid → magnesium chloride + hydrogen **[3]**

(e) sulphur dioxide + oxygen → sulphur trioxide **[3]**

(f) iron + bromine → iron(III) bromide **[3]**

(g) methane + oxygen → carbon dioxide + water **[3]**

(h) propan-1-ol + oxygen → carbon dioxide + water **[3]**

(i) silver carbonate → silver + carbon dioxide + oxygen **[3]**

(j) iron(III) oxide + hydrogen → iron + steam **[3]**

(k) calcium carbonate + hydrochloric acid → calcium chloride + water + carbon dioxide **[3]**

(l) copper(II) oxide + methane → copper + carbon dioxide + water **[3]**

(m) copper(II) oxide + ammonia → copper + nitrogen + water **[3]**

(n) ammonia + oxygen → nitrogen monoxide + water **[3]**

(o) potassium hydroxide + sulphuric acid → potassium sulphate + water **[3]**

(p) barium nitrate + sodium sulphate → barium sulphate + sodium nitrate **[3]**

(q) nitric acid + magnesium hydroxide → magnesium nitrate + water **[3]**

(r) potassium hydrogencarbonate → potassium carbonate + water + carbon dioxide **[3]**

(s) lead(II) nitrate + potassium iodide → lead(II) iodide + potassium nitrate **[3]**

(t) copper + nitric acid → copper(II) nitrate + nitrogen dioxide + water. **[3]**

1.16 Write fully balanced chemical equations for the following reactions (do not include state symbols):

(a) cyclopropene, C_3H_4, undergoing complete combustion **[3]**

(b) propene reacting with hydrogen **[3]**

(c) the cracking of one mol of decane, $C_{10}H_{22}$, to give two mol of ethene and one mol of another product **[3]**

(d) chlorine and propane, C_3H_8, in sunlight to give a disubstituted product **[3]**

(e) propane-1,2-diol, $C_3H_6(OH)_2$, reacting with sodium **[3]**

(f) buta-1,3-diene, C_4H_6, reacting completely with bromine [3]

(g) magnesium bromide reacting with silver nitrate to give a precipitate of silver bromide [3]

(h) potassium chloride and concentrated sulphuric acid [3]

(i) hydrogen iodide and concentrated sulphuric acid [3]

(j) the thermal decomposition of potassium chlorate(VII), $KClO_4$, to potassium chloride and oxygen [3]

(k) aluminium hydroxide, $Al(OH)_3$, and dilute nitric acid [3]

(l) lead(IV) oxide, PbO_2, and concentrated hydrochloric acid to give chlorine [3]

(m) the action of heat on lithium carbonate, Li_2CO_3 [3]

(n) ammonium dichromate(VI), $(NH_4)_2Cr_2O_7$, decomposing when heated to form chromium(III) oxide, nitrogen and steam [3]

(o) lead(II) ethanoate, $(CH_3CO_2)_2Pb$, reacting with potassium chromate(VI), K_2CrO_4, to give a precipitate of lead(II) chromate(VI). [3]

1.17 Write balanced ionic equations for the following reactions (include state symbols):

(a) dilute nitric acid and barium hydroxide solution, $Ba(OH)_2(aq)$ [3]

(b) the reaction of copper metal with silver nitrate solution [3]

(c) the reaction of thiosulphate ions, $S_2O_3^{2-}$, with iodine, I_2 [3]

(d) lithium carbonate, Li_2CO_3, and dilute sulphuric acid [3]

(e) the precipitation of lead(II) iodide, PbI_2 [3]

(f) the oxidation of acidified iron(II) sulphate solution by hydrogen peroxide solution [3]

(g) the conversion of sodium chromate(VI), $Na_2CrO_4(aq)$, to sodium dichromate(VI), $Na_2Cr_2O_7(aq)$, by concentrated sulphuric acid [3]

(h) the reaction of gaseous chlorine(IV) oxide, ClO_2, with alkali to give a solution containing the chlorite(V) ion, ClO_2^-, and releasing oxygen [3]

(i) the reduction of chromium(III) ions to chromium(II) ions by zinc and hydrochloric acid [3]

(j) the reduction of manganate(VII) ions, $MnO_4^-(aq)$, to manganate(VI) ions, $MnO_4^{2-}(aq)$, by concentrated aqueous alkali. [3]

1.18 In the following set of questions you are required to calculate either the mass of a reactant needed or the mass of a product formed. Assume complete conversion of reactants to products. Start by writing the equation for the reaction involved.

(a) The mass of calcium carbonate (limestone) that would have to be heated in order to produce 500 tonnes of calcium oxide (quicklime). [4]

(b) The amount of glucose that would be needed to produce 11.5 g of ethanol by fermentation. [4]

(c) 40 cm^3 of a 0.10 M solution of lead(II) nitrate are mixed with an excess of magnesium metal. What mass of lead would be obtained? [4]

(d) Magnesium reacts with hydrogen ions to form a salt and hydrogen gas. What mass of magnesium would have to be added to an excess of dilute hydrochloric acid in order to obtain 480 cm^3 of hydrogen gas? (Assume that under laboratory conditions, one mole of hydrogen molecules has a volume of 24 000 cm^3.) **[4]**

(e) The mass of titanium produced and the mass of sodium necessary for the reduction of 4.75 t of titanium(IV) chloride, $TiCl_4$. **[4]**

1.19 This question involves the calculation of yields in a range of chemical reactions.

(a) Butan-1-ol can be converted to 1-bromobutane by means of the reaction:

$$C_4H_9OH + NaBr + H_2SO_4 \rightarrow C_4H_9Br + NaHSO_4 + H_2O$$

A student started with 8.10 g of butan-1-ol and an excess of the other reagents. He obtained 10.5 g of 1-bromobutane. What was his percentage yield? **[2]**

(b) N-phenylethanamide can be made by the addition of ethanoic anhydride to phenylamine in the presence of glacial ethanoic acid. The essential reaction is:

$$(CH_3CO)_2O + C_6H_5NH_2 \rightarrow CH_3CONHC_6H_5 + CH_3CO_2H$$

In a typical experiment 6.0 cm^3 of ethanoic anhydride and 4.0 cm^3 of ethanoic acid were refluxed with 4.0 cm^3 of phenylamine. After purification 3.8 g of product was obtained.

(i) Which was the limiting reagent? Show your working. **[2]**

(ii) What was the percentage yield? **[2]**

(Densities in g cm^{-3}: ethanoic anhydride 1.08, phenylamine 1.02)

(c) Cyclohexanol can be converted to cyclohexene by dehydration with concentrated phosphoric acid.

$$C_6H_{11}OH \rightarrow C_6H_{10} + H_2O$$

The percentage yield for this experiment is, typically, approximately 80%. In order to be reasonably certain of obtaining 12.5 g of cyclohexene, what mass of cyclohexanol must be used initially? **[2]**

1.20 Calculate the number of moles present in each of the following solutions:

(a) 25.0 cm^3 of 0.1 M hydrochloric acid **[1]**

(b) 1.5 dm^3 of 2.5 M sodium hydroxide **[1]**

(c) 3.0 cm^3 of 2.0 M sulphuric acid **[1]**

(d) 20.0 cm^3 of 0.17 M barium hydroxide **[1]**

(e) 11.2 cm^3 of 0.5 M sodium carbonate solution **[1]**

(f) 0.5 dm^3 of 2.0 M potassium hydroxide **[1]**

(g) 0.5 cm^3 of 0.4 M nitric acid **[1]**

(h) 10.8 cm^3 of 0.005 M silver nitrate **[1]**

(i) 1750 cm^3 of 0.080 M ethanoic acid **[1]**

(j) 10.3 cm^3 of 0.05 M iodine solution. **[1]**

1.21 What is the molarity (concentration in mol dm^{-3}) of the following solutions?

(a) A solution of hydrochloric acid, $HCl(aq)$, containing 73.0 g of hydrogen chloride dissolved in 2 dm^3 of solution. [2]

(b) A solution of sulphuric acid, $H_sSO_4(aq)$, containing 4.90 g of sulphuric acid dissolved in 500 cm^3 of solution. [2]

(c) A solution of potassium hydroxide, $KOH(aq)$, containing 1.40 g of potassium hydroxide dissolved in 100 cm^3 of solution. [2]

(d) A solution of calcium hydroxide, $Ca(OH)_2(aq)$, containing 0.37 g of calcium hydroxide dissolved in 1.5 dm^3 of solution. [2]

(e) A solution of methanoic acid, $HCO_2H(aq)$, containing 0.46 g of methanoic acid dissolved in 200 cm^3 of solution. [2]

(f) A solution of propanone, $CH_3COCH_3(aq)$, containing 1.45 g of propanone dissolved in 50 cm^3 of solution. [2]

(g) A solution of copper(II) sulphate-5-water, $CuSO_4.\,5H_2O(aq)$, containing 7.50 g of blue copper sulphate crystals dissolved in 250 cm^3 of solution. [2]

(h) A solution of cobalt(II) chloride-6-water, $CoCl_2.\,6H_2O(aq)$, containing 0.952 g of the hydrated crystals dissolved in 40 cm^3 of solution. [2]

(i) A solution of ammonium iron(II) sulphate-6-water, $(NH_4)_2SO_4FeSO_4.\,6H_2O(aq)$, containing 1.96 g of the solid dissolved in 0.10 dm^3 of solution. [2]

(j) A solution of potassium dichromate(VI), $K_2Cr_2O_7(aq)$, containing 0.98 g of the solid dissolved in 30 cm^3 of solution. [2]

1.22 Calculate:

(a) the number of moles of hydroxide ions present in 25.0 cm^3 of 0.05 M calcium hydroxide, $Ca(OH)_2$ [1]

(b) the number of moles of aluminium ions present in 30.0 cm^3 of 0.50 M aluminium sulphate, $Al_2(SO_4)_3$ [1]

(c) the number of moles of sulphate ions present in 30.0 cm^3 of 0.50 M aluminium sulphate, $Al_2(SO_4)_3$ [1]

(d) the number of moles of hydrogen ions present in 40.0 cm^3 of 0.75 M phosphoric acid, H_3PO_4 [1]

(e) the number of moles of lead(II) ions in 1 dm^3 of a solution of lead(II) nitrate, $Pb(NO_3)_2$, which is 0.50 M with respect to nitrate ions. [1]

1.23 In each part of this question, calculate the mass of the named substance which would have to be weighed out to make the solution described.

(a) Sodium hydroxide, $NaOH$ – to make 100 cm^3 of a 0.10 M solution. [2]

(b) Silver nitrate, $AgNO_3$ – to make 1000 cm^3 of a 0.010 M solution. [2]

(c) Propanone, CH_3COCH_3 – to make 50 cm^3 of a 2.0 M solution. [2]

(d) Potassium carbonate, K_2CO_3 – to make 20.0 cm^3 of a 0.025 M solution. [2]

(e) Copper(II) sulphate-5-water, $CuSO_4.5H_2O$ – to make 5 dm^3 of a 0.020 M solution. [2]

1.24 10.0 cm^3 of a solution of potassium hydroxide was titrated with a 0.10 M solution of hydrochloric acid. 13.5 cm^3 of the acid was required for neutralization.
  (a) Write the equation for the reaction. **[1]**
  (b) How many moles of hydrochloric acid were used in the titration? **[1]**
  (c) How many moles of potassium hydroxide must therefore have been present in 10.0 cm^3 of solution? **[1]**
  (d) Calculate the concentration of the potassium hydroxide solution in mol dm^{-3}. **[1]**

1.25 Sodium ethanedioate, $Na_2C_2O_4$, can be made into a solution of an exact and reliable concentration and so can be used as a standard solution for checking the concentration of a solution containing sulphuric acid. The reaction is summarized by the equation:

$$Na_2C_2O_4 + H_2SO_4 \rightarrow Na_2SO_4 + (CO_2H)_2$$

In a particular experiment 10.0 cm^3 of a 0.50 M solution of sodium ethanedioate was found to react with 8.7 cm^3 of sulphuric acid. Find the concentration of the sulphuric acid. **[2]**

1.26 The equation for the reaction of sodium hydroxide solution with sulphuric acid is:

$$2NaOH(aq) + H_2SO_4(aq) \rightarrow Na_2SO_4(aq) + 2H_2O(l)$$

20.0 cm^3 of a solution of sodium hydroxide of concentration 0.050 M was pipetted into a conical flask and titrated with sulphuric acid. 24.2 cm^3 of acid was required for neutralization. Calculate the concentration of the acid in mol dm^{-3}. **[3]**

1.27 2.7 g of impure sodium carbonate crystals of formula $Na_2CO_3.10H_2O$ was dissolved to make 100 cm^3 of solution. 10.0 cm^3 portions were titrated against hydrochloric acid of concentration 0.12 M. 16.4 cm^3 of the acid were needed for neutralization.
  (a) Suggest a suitable indicator for this reaction. **[1]**
  (b) Write an equation for the reaction. **[1]**
  (c) Calculate the number of moles of acid used. **[1]**
  (d) What is the number of moles of sodium carbonate used in the titration? **[1]**
  (e) Calculate the concentration of the sodium carbonate solution in mol dm^{-3}. **[1]**
  (f) What mass of the crystals does this solution contain? **[1]**
  (g) What is the percentage purity of the crystals? **[1]**

States of matter and changes of state

2

In this section you will need to understand:

- the differences in the arrangement of the particles in solids, liquids and gases and how these differences affect their physical properties
- what is meant by 'Brownian motion'
- the nature of 'molecular' and 'giant' structures and how these give rise to differences in physical properties.

2.1 Summarise the differences in structure between solids, liquids and gases. **[6]**

2.2 The movement of molecules in liquids is sometimes compared with the movement of dancers in a crowded disco. With what would you compare the movement of molecules in **(i)** a solid and **(ii)** a gas? **[2]**

2.3 What is Brownian motion? It can be viewed in various ways, for example, by observing pollen grains in water and by using a smoke cell. How would the motions compare in the two cases and how would you explain any differences observed? **[6]**

2.4 Explain the following:
- **(a)** It is easy to push in the plunger of a sealed syringe containing a gas but difficult if it contains a liquid. **[2]**
- **(b)** The pressure exerted by a gas in a sealed container increases as the temperature is raised. **[2]**
- **(c)** A gas will move to fill any container into which it is put. **[2]**
- **(d)** Pure water boils at 83 °C on the top of Mont Blanc (4840 metres above sea level). **[2]**

2.5 The experiment shown below is set up. What will happen if the plunger is pushed in until the volume of gas is halved? **[1]**

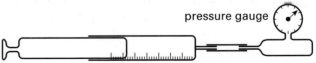

pressure gauge

Explain the observation in terms of how often and how hard the particles hit the wall. **[2]**

2.6 A strong sealed container containing ice was gently heated until only steam was present.
  (a) Draw a graph to show how the temperature inside the container changes with time.   [3]
  (b) Explain, in detail, its shape.   [10]

2.7 In the same group of the Periodic Table different elements and their compounds often have similar properties. Carbon and silicon are both in Group 4 but carbon dioxide, CO_2, has a boiling point of 195 K whereas silicon dioxide, SiO_2, has a boiling point of 1883 K. Explain this difference by a consideration of the structure and bonding in each compound.   [8]

2.8 How does the nature and arrangement of the particles in the six solids listed below explain the properties listed?

substance	melting point/K	boiling point/K	electrical conductivity of solid	
copper	1356	2840	conducts	[3]
diamond	3823	5100	does not conduct	[3]
iodine	387	457	does not conduct	[3]
lithium chloride	878	1613	does not conduct until it is melted	[3]
phenol	316	455	does not conduct	[3]

2.9 What conclusions can you reach about the type and arrangement of the particles which compose the substances whose properties are tabulated below?

substance	melting point/K	boiling point/K	electrical conductivity			
			of solid	of liquid	in water	
A	1933	3560	good	good	insoluble	[3]
B	161	166	poor	poor	insoluble	[3]
C	1043	1773	poor	good	good	[3]
D	203	331	poor	poor	good	[3]
E	179	322	poor	poor	insoluble	[3]
F	723	1263	poor	poor	insoluble	[3]
G	234	630	good	good	insoluble	[3]
H	3125	3873	poor	good	insoluble	[3]

3 Ionic bonding

In this section you will need to understand:

- how an ion differs from an atom
- which elements form positively charged ions and which form negatively charged ions
- the nature of an ionic bond
- how atoms become ions and vice versa
- the use of 'dot and cross' diagrams to show the formation of simple ionic substances
- the reasons for ionic compounds having certain characteristic properties
- the use of formulae for more complex ions
- how formulae are written for ionic compounds
- how simple experiments can be carried out to show that substances contain ions
- the arrangement of ions in lattice structures for simple ionic solids
- that there is no such thing as pure ionic bonding and why this is so.

3.1 **(a)** How is an ion different from an atom? **[1]**

(b) **(i)** What type of element forms positively charged ions? **[1]**

(ii) What type of element forms negatively charged ions? **[1]**

(c) What type of force exists in an ionic bond? **[1]**

(d) What particles are lost or gained when an atom turns into an ion? **[1]**

(e) Explain why ions of a certain charge are associated with particular groups of the Periodic Table. **[2]**

3.2 In this question you are only required to show the electron arrangements of atoms and ions in terms of the main energy levels.

(a) Write down the electron arrangement of a sodium atom. **[1]**

(b) Write down the electron arrangement of a sodium ion. **[1]**

(c) Write down the electron arrangement of a chlorine atom. **[1]**

(d) Write down the electron arrangement of a chlorine ion. **[1]**

(e) Show, using dot and cross diagrams, how sodium and chloride ions are formed from their respective atoms. **[2]**

(f) Indicate how the sodium and chloride ions which make up the structure of sodium chloride are arranged. **[3]**

3.3 Identify the numbers and symbols indicated by letters in the table below.

element	atomic number	number of electrons in ion	symbol for ion	
lithium	3	2	(a)	[1]
magnesium	12	(b)	Mg^{2+}	[1]
gallium	(c)	28	Ga^{3+}	[1]
cobalt(II)	27	(d)	Co^{2+}	[1]
cobalt(III)	27	(e)	Co^{3+}	[1]
silver	(f)	46	Ag^+	[1]
bromine	(g)	36	Br^-	[1]
oxygen	8	10	(h)	[1]
phosphorus	15	18	(i)	[1]
iodine	53	(j)	I^-	[1]

3.4 Write down the symbols for the following:
(a) a calcium ion [1]
(b) an aluminium ion [1]
(c) a sulphide ion [1]
(d) a copper(I) ion [1]
(e) a copper(II) ion [1]
(f) a nitride ion [1]
(g) a hydrogen ion [1]
(h) a hydride ion [1]
(i) a fluoride ion [1]
(j) a potassium ion. [1]

3.5 Draw dot and cross diagrams to indicate the formation of ions in:
(a) potassium chloride [2]
(b) calcium sulphide [2]
(c) sodium oxide [2]
(d) aluminium fluoride [2]
(e) magnesium nitride [2]
(f) magnesium chloride. [2]

3.6 Ionic compounds have certain characteristic properties. These include:
(i) relatively high melting and boiling points [4]
(ii) solubility in water [4]
(iii) insolubility in organic solvents [4]
(iv) the ability to conduct electricity when molten or in aqueous solution. [4]
Explain how these properties are related to their structures.

3.7 Some ions are composed of more than one element. Examples include the chromate(VI) ion, CrO_4^{2-}, and the sulphite ion, SO_3^{2-}. What are the formulae for the following ions?

(a) a nitrate ion [1]

(b) a nitrite ion [1]

(c) a phosphate(V) ion [1]

(d) a manganate(VII) ion [1]

(e) a chlorate(V) ion [1]

(f) a bromate(I) ion [1]

(g) a dichromate(VI) ion [1]

(h) a sulphate ion [1]

(i) a manganate(VI) ion [1]

(j) an iodate(VII) ion [1]

3.8 Write down the formulae for the following compounds:

(a) iron(II) bromide [1]

(b) iron(III) phosphate(V) [1]

(c) zinc sulphate [1]

(d) barium manganate(VII) [1]

(e) silver chromate(VI) [1]

(f) potassium bromate(I) [1]

(g) potassium bromate(III) [1]

(h) aluminium nitrate [1]

(i) sodium nitrite [1]

(j) sodium sulphite. [1]

3.9 Describe **two** simple experiments to show that ions move towards electrodes of opposite charge. [2×3]

3.10* Magnesium oxide has a structure which is described as two interpenetrating face-centred cubes, one of magnesium ions and one of oxide ions.

(a) Draw a diagram to illustrate this description. [4]

(b) (i) What is a 'unit cell'? [1]

(ii) Draw the unit cell for the magnesium oxide structure. [2]

(c) (i) What is meant by the term 'co-ordination number'? [2]

(ii) What are the co-ordination numbers of the two different ions in magnesium oxide? [2]

(d) Magnesium oxide has a similar structure to sodium chloride. How would you expect their melting and boiling points to compare? Explain your answer. [3]

(e) Name another substance which you would expect to have a similar solid crystalline lattice to that of sodium chloride. [1]

3.11* Unlike sodium chloride, calcium fluoride (fluorite) has a 1 : 2 ratio of metal : halide ions.

   (a) Draw a diagram to show how the ions are arranged in the crystalline lattice of calcium fluoride. **[4]**

   (b) What are the co-ordination numbers of the two ions in this case? **[2]**

3.12 The bonding between the ions of opposite charge in an ionic substance is never totally ionic.

   (a) Explain why this is so. **[3]**

   (b) What name is given to the feature you have described in **(a)**? **[1]**

   (c) What are some of the consequences of this lack of a pure bond type? **[2]**

   (d) How do the charges and sizes of the ions affect the degree of ionic bonding in a substance? **[4]**

   (e) How would you expect the bonding in beryllium iodide to compare with that in barium fluoride? Explain your answer. **[3]**

4 Energetics 1

In this section you will need to understand:

- what is meant by the terms 'enthalpy' and 'enthalpy change' and, specifically, what is meant by 'enthalpy change of reaction', 'enthalpy change of formation', 'enthalpy change of combustion' and 'enthalpy change of atomization'
- how enthalpy changes are represented when writing chemical equations
- the conditions which define 'standard' enthalpies and 'standard' enthalpy changes
- the use of 'energy level diagrams'
- Hess's law and its use in calculating unknown enthalpy changes by the use of Hess cycles
- the calculation of the energy change occurring when a known mass of water or a dilute solution changes in temperature
- what is meant by the term 'bond energy' and how average bond energies can be used to calculate the enthalpy change of atomization of a gaseous compound.

4.1 What is meant by the terms 'enthalpy' and 'enthalpy change'? **[2]**

4.2 The enthalpy change for a reaction is often indicated by adding to the right of a chemical equation an extra term of the form $\Delta H = \pm x$ kJ mol^{-1}. Explain the meaning of each symbol in this additional statement. **[4]**

4.3 Enthalpy changes are usually stated with reference to a standard set of conditions. What are these standard conditions? **[4]**

4.4 Draw energy level diagrams to indicate the following energy changes:
(i) $C_2H_4(g) + Br_2(l) \rightarrow C_2H_4Br_2(l)$; $\Delta H^{\ominus} = -133.0$ kJ mol^{-1} **[2]**
(ii) $N_2(g) + O_2(g) \rightarrow 2NO(g)$; $\Delta H^{\ominus} = +180.4$ kJ mol^{-1}. **[2]**

4.5 Explain the meaning of the terms:
(a) standard enthalpy change of formation **[3]**
(b) standard enthalpy change of combustion. **[3]**

Data: Data needed in some of the following questions:
$C(s) + O_2(g) \rightarrow CO_2(g)$; $\Delta H_f^{\ominus}[CO_2(g)] = -393.5$ kJ mol^{-1}
$H_2(g) + 0.5O_2(g) \rightarrow H_2O(l)$; $\Delta H_f^{\ominus}[H_2O(l)] = -285.8$ kJ mol^{-1}

4.6 **(a)** State Hess's law. [2]

(b) Use Hess's law to calculate the standard enthalpy changes of reaction, $\Delta H^{\ominus}$, in the following examples:

(i)

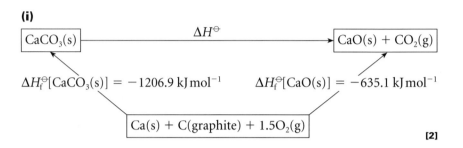

[2]

(ii)

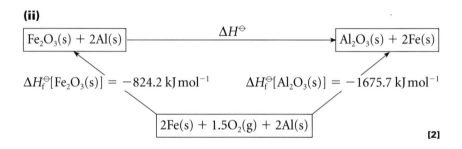

[2]

(iii)

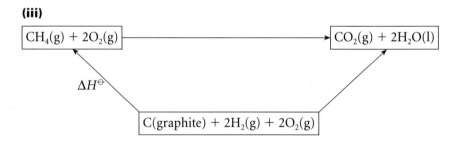

given that $\Delta H_c^{\ominus}[CH_4(g)] = -890.3 \text{ kJ mol}^{-1}$ [2]

What is the full name of the energy change which you have calculated in **(iii)**?

[2]

4.7 **(a)** Use the following data to construct a suitable Hess cycle for calculating the standard enthalpy change of the third reaction below. Use the cycle to show that the required enthalpy change is $-86.6 \text{ kJ mol}^{-1}$.

$$Zn(s) + CuO(s) \rightarrow ZnO(s) + Cu(s); \Delta H^{\ominus} = -192.8 \text{ kJ mol}^{-1}$$

$$Zn(s) + H_2O(g) \rightarrow ZnO(s) + H_2(g); \Delta H^{\ominus} = -106.2 \text{ kJ mol}^{-1}$$

$$CuO(s) + H_2(g) \rightarrow Cu(s) + H_2O(g); \Delta H^{\ominus} = \pm x \text{ kJ mol}^{-1}$$ [4]

(b) Set up and use a suitable Hess cycle in order to determine the standard enthalpy change for the reaction involving cyclobutane shown below:

$$C_4H_8(g) + 6O_2(g) \rightarrow 4CO_2(g) + 4H_2O(l)$$

(The standard enthalpy change of formation of cyclobutane = +3.7 kJ mol^{-1}.)

[4]

What is the full name of the enthalpy change calculated? [2]

Data: In the next few questions it will often be necessary to use the specific heat capacity of water which has the value of 4.18 J g^{-1} °C^{-1}

4.8 When water changes in temperature, the amount of energy transferred is found from the equation:

$$\text{Enthalpy change} = \left(\begin{array}{c} \text{specific heat capacity} \\ \text{of water} \end{array} \right) \times \left(\begin{array}{c} \text{mass of} \\ \text{water} \end{array} \right) \times \left(\begin{array}{c} \text{temperature} \\ \text{change} \end{array} \right)$$

A person placed his hand in 400 g of cold water for 15 minutes. A temperature change of 5.0 °C was recorded.

Calculate the enthalpy change of the water. [2]

4.9 When 0.32 g of calcium was added to 50 cm^3 of water, a temperature rise of 12.3 °C took place. On neutralizing the resulting solution of calcium hydroxide with 4.6 cm^3 of dilute hydrochloric acid, a further temperature rise of 3.2 °C occurred. When 0.29 g of calcium was added to 50 cm^3 of dilute hydrochloric acid (an excess), the temperature rose by 14.5 °C.

(a) Calculate the amount of energy given out in each of the three reactions. (Assume that all the solutions have the same specific heat capacity as water and have a density of 1.0 g cm^{-3}.) [6]

(b) What are the ΔH values of the corresponding enthalpy changes expressed in kJ mol^{-1}? [6]

(c) Write equations for the three reactions. [6]

(d) Combine the three equations to produce a Hess cycle. [2]

(e) Use your Hess cycle to show that within the limits of experimental error Hess's law is obeyed for this set of reactions. [4]

4.10 The standard enthalpy change of combustion of propane is −2219 kJ mol^{-1}.

(a) (i) Write the molecular formula for propane. [1]
   (ii) Write an equation for the complete combustion of propane. [2]

(b) What is the standard enthalpy change of propane per gram of propane? [2]

(c) Propane is used as a fuel. How much energy will be produced when 1 kg of propane is completely burnt? [1]

4.11 The standard enthalpy changes of combustion for a series of alcohols are shown in the following table:

alcohol	structural formula	standard enthalpy change of combustion, $\Delta H_c^{\ominus}$/kJ mol^{-1}
propan-2-ol	$CH_3CHOHCH_3$	-2006
butan-2-ol	$CH_3CHOHCH_2CH_3$	-2660
pentan-2-ol	$CH_3CHOHCH_2CH_2CH_3$	-3313
hexan-2-ol	$CH_3CHOHCH_2CH_2CH_2CH_3$	-3967
heptan-2-ol	$CH_3CHOHCH_2CH_2CH_2CH_2CH_5$	?

(a) Calculate the differences in value between the standard enthalpy changes for each successive pair of alcohols. **[2]**

(b) What do you notice about the differences in value which you have calculated in (a)? What is the explanation? **[3]**

(c) Estimate the standard enthalpy change of combustion of heptan-2-ol. **[2]**

(d) Write an equation for the complete combustion of heptan-2-ol and then use an energy level diagram to represent the change taking place. **[4]**

4.12 (a) Write an equation for the complete combustion of propene, C_3H_6, for which the standard enthalpy change is -2058.1 kJ mol^{-1}. **[2]**

(b) (i) Construct a Hess cycle from which the enthalpy change of formation of propene may be found. **[2]**

(ii) Use the cycle to determine the standard enthalpy change of formation of propene. **[2]**

(c) (i) By a similar method, determine the enthalpy change of formation of propan-1-ol, the enthalpy change of combustion of which is -2021.0 kJ mol^{-1}. **[2]**

(ii) Repeat the calculation for phenol, the enthalpy change of combustion of which is -3053.4 kJ mol^{-1}. **[2]**

4.13 Find the enthalpy change of combustion of phosphine, PH_3, given that its standard enthalpy change of formation is $+5.4$ kJ mol^{-1}. (The standard enthalpy change of formation of phosphorus(V) oxide, $P_4O_{10}(s)$ is -2984.0 kJ mol^{-1}. **[4]**

4.14 (a) (i) Buta-1,2-diene has the formula $CH_2{=}C{=}CHCH_3$. It is a gaseous compound. Using the average bond energies given below, determine the enthalpy change of atomization for the compound. **[3]**

$$E(C{-}H) = +413 \text{ kJ mol}^{-1}; E(C{=}C) = +612 \text{ kJ mol}^{-1}$$
$$E(C{-}C) = +347 \text{ kJ mol}^{-1}$$

(ii) Carry out a similar calculation to that in (a)(i) for buta-1,3-diene which has the formula $CH_2{=}CHCH{=}CH_2$. **[2]**

(b) Using the values below for the standard enthalpy changes of formation of the butadienes, together with the standard enthalpy changes of atomization of carbon and hydrogen, calculate the enthalpy changes of atomization for the two compounds. **[4]**

$$\Delta H_f^{\ominus}[CH_2\!=\!C\!=\!CH_2\!-\!CH_3(g)] = +162.3 \text{ kJ mol}^{-1}$$
$$\Delta H_f^{\ominus}[CH_2\!=\!CHCH\!=\!CH_2(g)] = +109.9 \text{ kJ mol}^{-1}$$

$$\Delta H_{at}^{\ominus}[C(graphite)] = +716.7 \text{ kJ mol}^{-1} \text{ of atoms formed}$$
$$\Delta H_{at}^{\ominus}[0.5H_2(g)] = +218.0 \text{ kJ mol}^{-1} \text{ of atoms formed.}$$

(c) Compare the values you have obtained for the enthalpy changes of atomization of the butadienes in **(a)(i)**, **(a)(ii)** and **(b)** and suggest a reason for any differences obtained. **[2]**

5 Kinetics 1

In this section, you will need to understand:

- the meaning of the term 'rate of reaction'
- that reactions can vary in rate from very fast to very slow
- which four factors affect the rate of a chemical reaction and why they do so
- the principles of 'collision theory'
- the changes that take place in bonding and in energy when a chemical reaction occurs
- what is meant by the term 'activation energy' and the effect that catalysts have on its value
- what is meant by the term 'energy profile' for a chemical reaction
- the essential difference between homogeneous catalysis and heterogeneous catalysis
- how enzymes differ in their action compared with conventional catalysts.

5.1 **(a)** What do you understand by the term 'rate of reaction'? **[2]**

(b) State **two** reactions which take place very rapidly. **[2]**

(c) State **two** reactions which take place very slowly. **[2]**

5.2 **(a)** Write down **four** factors which can have a significant effect on the rate of a chemical reaction. **[4]**

(b) State briefly how changes in each of the four factors you have given in your answer to **(a)** usually affect the rate of a reaction. **[4]**

5.3 **(a)** The way in which chemicals react together is often described in terms of 'the collision theory'. What is meant by 'the collision theory'? **[1]**

(b) Describe how the collision theory can be used to explain the effect of the factors you have given in your answer to **5.2(a)**. **[8]**

5.4 **(a)** What changes in chemical bonding take place when a reaction occurs? What types of energy change accompany these changes in bonding? **[4]**

(b) Although particles of reactants have to meet before a chemical reaction can occur, only a small proportion of the collisions which take place are *successful* in terms of producing chemical change. Give **two** factors which account for this lack of success. **[2]**

(c) How does the simplest form of collision theory therefore have to be modified when applied to 'real' reactants? **[4]**

(d) Explain why raising the temperature increases the proportion of successful collisions between reacting particles. **[2]**

5.5 **(a)** What is a 'catalyst'? [2]

(b) Describe briefly the main features associated with catalytic function. [8]

(c) Explain why a catalyst is able to alter the rate at which products are formed during a chemical reaction. [4]

5.6 The activation energy for the reaction:

$$N_2(g) + 3H_2(g) \rightarrow 2NH_3(g); \Delta H^\ominus = -92.2 \text{ kJ mol}^{-1}$$

is $+335 \text{ kJ mol}^{-1}$.

However, when carried out in the presence of tungsten, the value of the activation energy is $+60 \text{ kJ mol}^{-1}$.

(a) How is the tungsten acting? Explain your answer. [3]

(b) Sketch an energy profile to display the information given above. [5]

5.7 **(a)** Catalysts are divided into *homogeneous* and *heterogeneous*. Explain the meanings of the two italicized words. [2]

(b) Give **two** examples in each case of chemical reactions in which **(i)** a homogeneous catalyst is used and **(ii)** a heterogeneous catalyst is used. [4]

5.8 Enzymes are catalysts present in living systems. State how enzymes are:
(a) similar to, **(b)** different from conventional catalysts.
(You should mention **six** comparisons in total in order to obtain full marks.) [6]

6 Equilibrium 1

- the meaning of the terms 'reversible reaction', 'forward reaction' and 'backward reaction'
- what is meant by 'a state of equilibrium' and what its characteristics are
- how chemical equilibrium can be demonstrated to be dynamic in nature
- the effect of a catalyst on the position and rate of attainment of equilibrium
- the application and use of Le Chatelier's principle
- qualitatively, the reasons for the choice of operating conditions for a reversible industrial process such as the Haber Process.

6.1 **(a)** A reaction is said to be 'reversible'. What do you understand by the term 'reversible reaction'? **[1]**

(b) Explain what would eventually happen if reactants for a reaction which is reversible were placed in a sealed container. **[2]**

6.2 A saturated solution of potassium iodide in contact with solid potassium iodide has reached a state of equilibrium.

(a) State **four** characteristics of a state of equilibrium. **[4]**

(b) What change, if any, would take place if more solid potassium iodide were to be added to the mixture, the temperature being kept constant? **[1]**

(c) If the added potassium iodide were labelled with a radioactive isotope of iodine, ^{131}I, (half-life 8.7 days) what observation can be made after a few hours? What deduction could be made from this observation? **[3]**

6.3 The equation for a chemical reaction at equilibrium could be of the type:

$$A + B \rightleftharpoons C + D.$$

(a) What, if anything, can you say about the amounts of A, B, C and D present at equilibrium? **[1]**

(b) What, if anything, can you say about the rates of the forward and backward reactions when equilibrium has been reached? **[1]**

(c) What, if anything, can you say about the enthalpy changes of reaction for the forward and backward reactions? **[2]**

(d) If a catalyst is added to the reaction mixture, what effect would this have on the rate of attainment and the position of equilibrium? **[2]**

6.4 The Haber Process, used for the industrial production of ammonia, is based upon the reaction:

$$N_2(g) + 3H_2(g) \rightleftharpoons 2NH_3(g); \Delta H = -92.2 \text{ kJ mol}^{-1}.$$

The graph below shows how the amount of ammonia present at equilibrium varies with the pressure and temperature used.

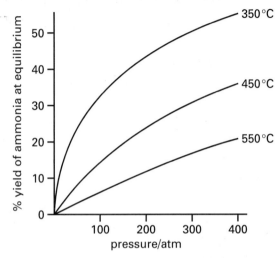

(a) (i)   From the graph, what conclusion can you reach about the effect of temperature on the position of equilibrium of this reaction?   [1]
(ii) What type of energy change takes place when ammonia is formed from nitrogen and hydrogen?   [1]
(b) (i)   From the graph, what conclusion can you reach about the effect of pressure on the position of equilibrium of this reaction?   [1]
(ii) How does the number of molecules present change when ammonia is formed from nitrogen and hydrogen?   [1]
(c) From many similar studies, Le Chatelier summarized the way that equilibrium reactions behave when subjected to a change of conditions. Using the principle that he formulated, explain the conclusions which you have reached in (a) parts (i) and (ii) and (b) parts (i) and (ii).   [5]
(d) The conditions used in the Haber Process are, typically, 200 atmospheres and 450 °C, together with a catalyst. Why do you think such conditions are favoured? Explain your answer.   [5]

6.5 (a) How will an increase in pressure affect the amount of products obtained in each of the following reactions? Explain how you reach your conclusion in each case.
(i) $PCl_5(g) \rightleftharpoons PCl_3(g) + Cl_2(g)$ [2]
(ii) $CH_4(g) + H_2O(g) \rightleftharpoons CO(g) + 3H_2(g)$ [2]
(iii) $N_2(g) + O_2(g) \rightleftharpoons 2NO(g)$ [2]
(iv) $2SO_2(g) + O_2(g) \rightleftharpoons 2SO_3(g)$ [2]

(b) How will an increase in temperature affect the amount of products obtained in each of the following reactions?

(i) $2H_2(g) + O_2(g) \rightleftharpoons 2H_2O(g)$ exothermic in forward direction [1]

(ii) $PCl_5(g) \rightleftharpoons PCl_3(g) + Cl_2(g)$ endothermic in forward direction [1]

(iii) $CH_3CO_2H(l) + C_2H_5OH(l) \rightleftharpoons CH_3CO_2C_2H_5(l) + H_2O(l)$ almost no energy change [1]

(iv) $C(s) + CO_2(g) \rightleftharpoons 2CO(g)$ endothermic in forward direction. [1]

7 The structure of the atom

In this section you will need to understand:

- the relationship between the numbers of protons and electrons in an atom
- the structure of the nucleus
- the fact that atoms of the same element may contain different numbers of neutrons
- how a mass spectrometer can be used to determine relative atomic masses
- how mass spectrometry can be used to determine the relative abundance of isotopes.

7.1 This question is concerned with the particles found in atoms.
  (a) Name the atomic particle
    (i) which has no charge [1]
    (ii) which has the lowest mass. [1]
  (b) Name the two atomic particles found in the nucleus. [2]
  (c) Where are the electrons in an atom and how are they held in place? [2]

7.2 **(a) (i)** What is meant by the term 'atomic number'? [1]
    (ii) What information does it give about the number of *electrons* in an atom? [1]
  (b) State the number of protons and electrons in the atom or ion represented by the following symbols:
    He, Na, Na^+, O, O^{2-}, F, F^-, Al, Al^{3+}, N, N^{3-}. [11]

7.3 **(a)** Why are there always equal numbers of protons and electrons in an atom? [2]
  (b) Sodium atoms lose one electron in forming sodium ions. Explain why these ions have a positive charge. [2]
  (c) What is meant by the term 'mass number'? [1]
  (d) Knowing the mass number of an element, what other information is needed to work out the number of neutrons in an atom of that element? [1]

7.4 The element boron has an atomic number of 5 and has atoms with mass numbers 10 and 11.
  (a) What name is given to atoms of boron with different mass numbers? [1]
  (b) Write down the conventional symbols for these two atoms, showing their atomic and mass numbers. [2]
  (c) Boron contains 18.7% of atoms with mass number 10 and 81.3% of atoms with mass number 11. Calculate the relative atomic mass of the element boron. [2]
  (d) Sketch the expected mass spectrum of boron. [3]

7.5 Magnesium has three isotopes with the relative isotopic masses and percentage abundances shown: ^{24}Mg 78.6%; ^{25}Mg 10.1%; ^{26}Mg 11.3%

  (a) Calculate the relative atomic mass of magnesium to three significant figures. [2]

  (b) Sketch the expected mass spectrum of magnesium. [3]

7.6 Write down the correct answers corresponding to **(a)**–**(m)** in the table below.

symbol	atomic number of element	mass number of element	number of protons	number of electrons	number of neutrons	overall charge	
(a)	**(b)**	**(c)**	**(d)**	10	12	0	[4]
^{7}Li$^+$	3	**(e)**	**(f)**	**(g)**	**(h)**	**(i)**	[5]
(j)	**(k)**	27	13	**(l)**	**(m)**	+3	[4]

7.7 Write down the correct answers corresponding to **(a)**–**(m)** in the table below.

symbol	atomic number	mass number	number of protons	number of electrons	number of neutrons	overall charge	
(a)	1	**(b)**	**(c)**	2	0	**(d)**	[4]
(e)	17	**(f)**	**(g)**	18	18	**(h)**	[4]
^{18}O^{2-}	8	**(i)**	**(j)**	**(k)**	**(l)**	**(m)**	[5]

7.8 The table below contains information about isotopes of the element silicon.

isotope	relative abundance (%)
$^{28}_{14}$Si	92.3
$^{29}_{14}$Si	4.6
$^{30}_{14}$Si	3.1

Explain why each of the following statements concerning silicon is correct:

  (a) silicon has a relative atomic mass between 28.0 and 28.5 [2]

  (b) all atoms of silicon contain 14 electrons [1]

  (c) atoms of silicon may contain 14, 15 or 16 neutrons. [2]

The electronic
8 configuration of atoms

In this section you will need to understand:

- the concept of energy levels in an atom
- the further subdivision of energy levels into s, p and d atomic orbitals
- the order of filling of the s, p and d atomic orbitals
- how to work out the electronic structure of an atom from a knowledge of its atomic number
- the meaning of the terms s-, p- and d-block element.

8.1 **(a)** The electrons in atoms are arranged in energy levels. Which energy level has the highest energy: the one nearest the nucleus, or the one furthest away from it? [1]

(b) How many electrons do the 1st, 2nd and 3rd energy levels contain when they are full? [3]

8.2 **(a)** Draw diagrams to show the arrangement of the electrons in energy levels of the following elements:
N, Na, Ne, S, Ca. [5]

(b) What do you notice about the energy levels in the neon atom? [1]

(c) What is the consequence of this arrangement on the chemical behaviour of neon? [2]

8.3 **(a)** Which of the following atoms have unpaired electrons?
He, K, N, Cl, Al, Mg [4]

(b) Which of the following ions have unpaired electrons?
Ca^{2+}, Ti^{2+}, Sc^{3+}, Cu^{+}, Mn^{3+}, Zn^{2+} [2]

8.4 Energy levels are further subdivided into atomic orbitals, represented by the letters s, p and d. What is the maximum number of electrons which each of these orbitals can hold? [3]

8.5 The element carbon has atomic number 6 and its electronic structure can be represented as $1s^2\, 2s^2\, 2p^2$. Use the same method to represent the arrangement of the electrons in the atomic orbitals of the following elements:
N, Na, Ne, S, Ca. [5]

8.6 **(a)** Write down the electronic structures of the elements having the following atomic numbers:

5, 8, 12, 9, 14, 18. [6]

(b) Write down the electronic structures of the following ions:

Li^+, S^{2-}, Cl^-, Al^{3+}, Ca^{2+}, K^+. [6]

8.7 Refer to a copy of the Periodic Table and write down the symbols for:

(a) (i) a positive ion containing the following numbers of electrons:

2, 10, 18, [3]

(ii) a negative ion containing the following numbers of electrons:

2, 10, 18. [3]

(b) Write down the symbol for a positive ion which has the same number of electrons as each of the following:

Ne, He, F^-, S^{2-}, Ar, C^{4-}. [6]

8.8 Write the symbol for a negative ion which has the same number of electrons as each of the following:

Na^+, Mg^{2+}, Ar, K^+, He. [5]

8.9 An element is described as being an s-, p-, or d-block element depending on the orbital occupied by the last electron added to the atom of the element. Classify each of the following elements as s-, p-, or d-block elements:

argon, calcium, carbon, iron, magnesium, scandium, sodium, vanadium. [8]

Electronegativity and its application to bond type

9

In this section you will need to understand:

- the definition of electronegativity
- how differences in electronegativity lead to polarity in covalent bonds
- the fact that some molecules containing polar bonds are themselves polar and attract each other as a result
- hydrogen bonds involving N—H, O—H and F—H bonds
- van der Waals' forces that are present in all matter, but are much weaker than hydrogen bonds
- the link between the physical properties of a substance and the intermolecular forces present.

9.1 What is meant by the term 'electronegativity'? [2]

9.2 The table shows the electronegativities of some elements, as indicated on the Pauling Scale.

element	H	B	C	N	O	F	S	Cl
electronegativity	2.1	2.0	2.5	3.0	3.5	4.0	2.5	3.0

(a) Choose, in each case, **two** of these elements which would form a covalent bond which was:

(i) non-polar,

(ii) most polar. [2]

(b) (i) Which **two** elements would form the most polar bonds with hydrogen? [2]

(ii) What name is given to the intermolecular forces between molecules which contain highly-polarised bonds involving hydrogen and another element? [1]

(iii) Give **two** physical properties which are strongly affected by the existence of such intermolecular forces. [2]

(c) Polarity in covalent bonds can be represented as follows, $H^{\delta+}$—$F^{\delta-}$, where F is the more electronegative element. Refer to the above table of electronegativities and indicate the polarity of the following bonds in the same way:

B—H, C—O, Cl—F, S—H, C—F, N—H. [6]

9.3 Electronegativity depends on the atomic number of an element. Electronegativity decreases with increasing atomic number down a group, but increases with increasing atomic number across a period. Explain these two trends. [4]

9.4 Some molecules which contain polar bonds are not themselves polar because they are symmetrical. Give the shapes of the following molecules and decide which of them would **not** be polar:
CF_4, NH_3, CO_2, H_2O, SO_2, CH_3Cl, $BeCl_2$, XeF_4. [12]

9.5 Arrange the following three substances in order of increasing boiling point, explaining your choice:
ethanol, ethanoic acid and ethanal. [5]

9.6 Explain the difference in the boiling points of ethanol, CH_3CH_2OH, (78 °C) and methoxymethane, CH_3OCH_3, (−23 °C). [3]

9.7 Plot the boiling points (in °C) of the following hydrides against the atomic number of the other element which they contain:

hydride	H_2O	H_2S	H_2Se	H_2Te
boiling point/°C	100	−61	−42	−2

Explain the shape of your plot. [8]

9.8 **(a)** Carbon dioxide is a gas at room temperature, while silicon dioxide is a solid with a boiling point in excess of 2000 °C. Explain this difference in boiling point in terms of their structures. [4]

(b) Explain why graphite is a soft material which conducts electricity well, while diamond is very hard and an electrical insulator. [4]

The factors influencing the shapes of molecules and ions

10

In this section you will need to understand:

- how repulsion between electron pairs makes them move as far apart as possible
- how molecular shape is determined by the number of bond pairs and lone pairs surrounding the central atom.

10.1 For each of the following molecules, give the number of bond pairs and the number of lone pairs: CH_4, NH_3, $BeCl_2$, H_2O, BF_3, SF_4. **[12]**

10.2 For each of the following ions NH_4^+, BH_4^-, CH_3^+, PCl_4^+, PCl_6^-, ICl_4^-, state:
(a) the number of bond pairs and the number of lone pairs **[12]**
(b) the expected shape of the ion. **[6]**

10.3 Bond pairs (BP) and lone pairs (LP) repel each other. Between which is the repulsion **(a)** greatest and **(b)** least? Suggest an explanation. **[6]**

10.4 Draw dot and cross diagrams of each of the following. Use your diagrams to explain which of them has bond angles of 120°. Show only outer electron shells. NH_3, BF_3, PCl_3. **[7]**

10.5 Sketch, and state the shape of each of the following: CH_3^+, H_3O^+, ICl_2^+, ClO_3^-. **[6]**

10.6* How can you tell from the positions of the elements in the Periodic Table that BH_4^-, CH_4 and NH_4^+ must have the same shape? Using similar reasoning, write the formula of a molecule having the same shape as H_3O^+. **[5]**

10.7 Explain the bond angles in the following species:
CH_4 109.5°, NH_3 107.3°, H_2O 104.5°. **[4]**

10.8 The bond angles in ammonia (NH_3) and phosphine (PH_3) are 107.3° and 94°, respectively. Suggest an explanation for this difference in bond angle. **[3]**

10.9* The shape of some molecules is influenced by considerations other than simply the number of lone and bond pairs, most often the formation of π-bonds. Use this information to explain the fact that trimethylamine, $(CH_3)_3N$, is pyramidal as expected but trisilylamine (in which silicon replaces carbon), $(SiH_3)_3N$, has a trigonal planar structure. **[5]**

10.10* Xenon hexafluoride, XeF_6, might be assumed to be octahedral in shape. Apply your knowledge of the factors which influence molecular shape to show that this assumption is incorrect. **[4]**

11 The concept of oxidation state and its uses

11.1 **(a)** Give the oxidation state of oxygen in the following species:
O_2, O^{2-}, O_2^{2-}. [3]

(b) Give the oxidation state of the element other than oxygen present in each of the following oxides:
CO, CO_2, MgO, Al_2O_3, B_2O_3, Cl_2O_7, SO_3, P_4O_{10}. [8]

11.2 **(a)** Arrange the following fluorides in order of increasing oxidation state of the other element which they contain, giving the oxidation state in each case:
CF_4, BF_3, ClF_5, OF_2, SF_6, IF_7, HF. [7]

(b) Give the oxidation state of nitrogen in each of the following:
N_2, NH_3, N_2H_4, NH_4^+, NF_3, N^{3-}. [6]

(c) Arrange the following in order of increasing oxidation state of nitrogen, giving the oxidation state in each case:
N_2O_4, NO, N_2O, N_2O_5, NO_2, N_2O_3. [6]

11.3 What is the oxidation state of sulphur in each of the following oxoanions?
SO_3^{2-}, SO_4^{2-}, $S_2O_3^{2-}$, $S_4O_6^{2-}$ [4]

11.4 What increase in the oxidation state of the nitrogen atom occurs in each of the following changes?
(a) $NO_2 \rightarrow NO_3^-$ [1]
(b) $N_2 \rightarrow NO$ [1]
(c) $N_2O \rightarrow N_2O_4$ [1]
(d) $NO \rightarrow N_2O_4$ [1]

11.5 Make a list of the following changes under three headings: reductions, oxidations and those in which the oxidation state is unchanged. Give the change in oxidation state (with its sign) in each case.

$NO_2 \rightarrow N_2O_4$ $N_2 \rightarrow NH_3$ $NH_3 \rightarrow NO_3^-$ $N_2H_4 \rightarrow N_2$
$NH_3 \rightarrow NH_4^+$ $SO_4^{2-} \rightarrow SO_3^{2-}$ $SO_3 \rightarrow SO_3^{2-}$ $SO_2 \rightarrow SO_3^{2-}$
$H_2S \rightarrow SO_2$ $S_2O_3^{2-} \rightarrow SO_4^{2-}$ $H_2O \rightarrow H_3O^+$ $H_2O \rightarrow H_2O_2$ [20]

11.6 Write the number of electrons needed to balance the following half-reactions:

    (a) $Ca^{2+}(aq) + e^- \rightarrow Ca(s)$ [1]

    (b) $2I^-(aq) - e^- \rightarrow I_2(aq)$ [1]

    (c) $2Al^{3+}(aq) + e^- \rightarrow 2Al(s)$ [1]

    (d) $S^{2-}(aq) - e^- \rightarrow S(s)$.   [1]

11.7 Rewrite the following equations, adding the number of electrons required to balance each one:

    (a) $NO_3^- + 2H^+ + e^- \rightarrow NO_2 + H_2O$ [1]

    (b) $S + 2H^+ + e^- \rightarrow H_2S$ [1]

    (c) $O_2 + 4H^+ + e^- \rightarrow 2H_2O$ [1]

    (d) $PbO_2 + 4H^+ + e^- \rightarrow Pb^{2+} + 2H_2O$.   [1]

11.8 By adding H^+, H_2O and e^- as necessary, write balanced half-equations for the following reductions:

    (a) $SO_4^{2-} \rightarrow S$ [2]

    (b) $NO_3^- \rightarrow NH_3$ [2]

    (c) $BrO_3^- \rightarrow Br_2$ [2]

    (d) $ClO^- \rightarrow Cl_2$.   [2]

11.9 By adding H^+, H_2O and e^- as necessary, write balanced half-equations for the following conversions:

    (a) $SO_4^{2-} \rightarrow H_2SO_3$ [2]

    (b) $NO_3^- \rightarrow NO$ [2]

    (c) $N_2 \rightarrow NH_4^+$ [2]

    (d) $H_2O_2 \rightarrow H_2O$.   [2]

11.10* Which of the following species can, in theory, undergo disproportionation? Explain why the other species cannot disproportionate.

    SO_4^{2-}, N_2, NO_3^-, NH_3, CO, Cl^-, S^{2-}, NH_4^+, H_2O_2 [6]

11.11* Which of the following represent disproportionation reactions?

    $ClO^- \rightarrow Cl_2$ and Cl^-

    $BrO_3^- \rightarrow Br^-$ and BrO_4^-

    $SO_3^{2-} \rightarrow S$ and SO_4^{2-} [2]

11.12* Write a balanced redox equation in each case to represent the disproportionation of:

    (a) IO^- into I^- and IO_3^- [3]

    (b) ClO_3^- into Cl^- and ClO_4^- [3]

    (c) NO into N_2O and NO_2 [3]

    (d) NO_2^- into NO_3^- and NO [3]

    (e) S into SO_2 and S^{2-} [3]

    (f) N_2O into NH_3 and NO_2.   [3]

11.13* On heating, ammonium nitrite, NH_4NO_2, gives water and one product containing nitrogen. Use the concept of oxidation state to identify this product and to write a balanced equation for the reaction which occurs on heating. **[6]**

11.14* Ammonium nitrate, NH_4NO_3, also gives water and one product containing nitrogen on heating. Use the concept of oxidation state to identify this product and to write a balanced equation for the reaction which occurs. **[6]**

Data: Use the following half-equations to answer questions 15 to 17 below:

$$MnO_4^-(aq) + 8H^+(aq) + 5e^- \rightarrow Mn^{2+}(aq) + 4H_2O(l)$$
$$Fe^{3+}(aq) + e^- \rightarrow Fe^{2+}(aq)$$
$$Sn^{4+}(aq) + 2e^- \rightarrow Sn^{2+}(aq)$$
$$2CO_2(g) + 2H^+(aq) + 2e^- \rightarrow H_2C_2O_4(aq)$$
$$Cr_2O_7^{2-}(aq) + 14H^+(aq) + 6e^- \rightarrow 2Cr^{3+}(aq) + 7H_2O(l)$$

11.15 What volume of 0.1 M tin(II) chloride solution would react exactly with the following?
(a) 50.0 cm^3 of 0.1 M iron(III) sulphate solution **[3]**
(b) 25.0 cm^3 of 0.1 M potassium dichromate(VI) solution **[3]**
(c) 10.0 cm^3 of 0.05 M potassium manganate(VII) **[3]**

11.16 Calculate the volume of 0.05 M potassium manganate(VII) which will react completely with:
(a) 25.0 cm^3 of 0.5 M iron(II) sulphate solution **[3]**
(b) 20.0 cm^3 of 0.1 M tin(II) chloride solution **[3]**
(c) 30.0 cm^3 of 0.2 M ethanedioic acid solution **[3]**
(d) 25.0 cm^3 of 0.05 M chromium(III) sulphate solution. **[3]**

11.17 Calculate the molarity of a solution of potassium dichromate(VI) 20.0 cm^3 of which reacts exactly with each of the following:
(a) 20.0 cm^3 of 1.0 M iron(II) sulphate solution **[3]**
(b) 25.0 cm^3 of 0.2 M ethanedioic acid solution **[3]**
(c) 40.0 cm^3 of 0.05 M tin(II) chloride solution. **[3]**

11.18 1.50 g of iron wire was dissolved in sulphuric acid to give a solution of $Fe^{2+}(aq)$ and the solution made up to 250 cm^3. A 10 cm^3 portion of this solution was titrated with 0.01 M potassium manganate(VII) (permanganate) solution, 19.5 cm^3 being required to reach an end point.
$$5Fe^{2+}(aq) + MnO_4^-(aq) + 8H^+(aq) \rightarrow 5Fe^{3+}(aq) + Mn^{2+}(aq) + 4H_2O(l)$$
(a) (i) Calculate the number of moles of potassium manganate(VII) used in each titration.
(ii) Calculate the number of moles of $Fe^{2+}(aq)$ ions present in 10 cm^3 of the solution made from the iron wire. **[3]**

(b) Calculate the mass of iron present in 10 cm³ of the solution made from the iron wire. [2]

(c) Calculate the mass of iron present in the sample of iron wire. [2]

(d) Calculate the percentage of iron in the sample of iron wire. [2]

11.19 60 dm³ of air containing a trace of chlorine was passed through an excess of potassium iodide solution at room temperature and pressure, forming iodine and chloride ions.

$$Cl_2(g) + 2I^-(aq) \rightarrow 2Cl^-(aq) + I_2(aq)$$

The resulting solution was made up to 250 cm³ and 25.0 cm³ of it then required 20.6 cm³ of sodium thiosulphate solution, $Na_2S_2O_3$, of concentration 0.10 mol dm⁻³ to react completely with the iodine present.

$$2S_2O_3{}^{2-}(aq) + I_2(aq) \rightarrow 2I^-(aq) + S_4O_6{}^{2-}(aq)$$

(Molar volume of a gas at room temperature and pressure = 24 dm³)

(a) Calculate the number of moles of sodium thiosulphate used in each titration. [2]

(b) Calculate the number of moles of iodine present in 25.0 cm³ of solution. [2]

(c) Calculate the number of moles of iodine formed by reaction with the 60 dm³ sample of air. [2]

(d) Calculate the number of moles of chlorine present in the 60 dm³ sample of air. [2]

(e) Calculate the volume of chlorine present in the 60 dm³ sample of air. [2]

(f) Calculate the percentage by volume of chlorine in the sample of air. [2]

11.20 5.0 g of hydrated copper(II) sulphate, $CuSO_4.nH_2O$, was dissolved in water and the solution made up to 100 cm³. Excess potassium iodide was added, forming iodine according to the equation:

$$2Cu^{2+}(aq) + 4I^-(aq) \rightarrow 2CuI(s) + I_2(aq)$$

10.0 cm³ portions of the resulting solution containing iodine were titrated with 0.10 M sodium thiosulphate solution, 20.0 cm³ being required to react completely.

$$2S_2O_3{}^{2-}(aq) + I_2(aq) \rightarrow 2I^-(aq) + S_4O_6{}^{2-}(aq)$$

(a) Calculate the number of moles of sodium thiosulphate used in each titration. [1]

(b) Calculate the number of moles of iodine molecules, I_2, formed by the reaction between the copper ions in the 5.0 g sample of hydrated copper(II) sulphate and the potassium iodide. [2]

(c) Calculate the number of moles of copper ions in the 5.0 g sample of hydrated copper(II) sulphate. [1]

(d) Calculate the number of moles of water of crystallization, n, in the hydrated copper(II) sulphate. [3]

12 Atomic structure and the Periodic Table

In this section you will need to understand:

- how the filling of electron orbitals occurs and how it leads to the arrangement of elements in groups
- the fact that elements in the same group have the same outer electronic configuration
- the fact that elements with the same outer electronic configuration show very similar patterns of reactivity.

12.1 Write down the electronic structure of the following atoms using the s, p, d notation (e.g. potassium's electronic structure would be represented as $1s^2\, 2s^2\, 2p^6\, 3s^2\, 3p^6\, 4s^1$): H, Na, Al, N, Ca, Cl. **[6]**

12.2 Write down the electronic structure of the following ions using the s, p, d notation: Li^+, Mg^{2+}, O^{2-}, F^-, P^{3-}, Si^{4-}. **[6]**

12.3 Write down the electronic structure of the elements having the following atomic numbers using the s, p, d notation: 8, 9, 11, 13, 16, 17. **[6]**

12.4 Write down the electronic structure of the *ions* which would be formed by the elements in the previous question, using the s, p, d notation, bearing in mind that the ions all have noble gas electron structures. **[6]**

12.5 Which of the following pairs of elements are in the same group in the Periodic Table: S and Si, He and Ar, N and P, Ca and Al, Be and Mg, Cl and O? **[3]**

12.6 The atomic numbers of several pairs of elements are shown below. Which represent pairs of elements in the same group in the Periodic Table?
3 and 9, 10 and 18, 4 and 20, 5 and 15, 6 and 14, 9 and 17.
Can you see any pattern in the atomic numbers of elements in the same group? **[6]**

12.7* In the early form of the Periodic Table the elements were arranged in order of increasing relative atomic mass, rather than increasing atomic number, as in the modern form of the Periodic Table. Discuss the advantages of using atomic numbers rather than relative atomic masses. **[6]**

12.8* Lithium and magnesium form ions of similar size (Li^+ 0.060 nm, Mg^{2+} 0.065 nm). Suggest an explanation in terms of their positions in the Periodic Table. **[4]**

Reactions of the elements and compounds of Groups 2 and 7

13.1 **(a)** Arrange the following ions in order of increasing ionic radius: Ba^{2+}, Mg^{2+}, Sr^{2+}, Ca^{2+}. **[2]**

(b) State and explain the changes in atomic radius which occur with increase in atomic number in Group 2. **[3]**

(c) Which of the Group 2 elements would you expect to react most vigorously with water, and why? **[5]**

13.2 **(a)** Write a balanced equation, with state symbols, for the reaction between:
  (i) calcium and oxygen **[2]**
  (ii) calcium oxide and water **[2]**
  (iii) calcium hydroxide solution and carbon dioxide. **[2]**

(b) Despite the greater first and second ionization energies of magnesium, it appears to burn much more vigorously in air than barium does. Suggest an explanation. **[4]**

13.3 The melting points of the oxides of four Group 2 metals are (but not in this order) 2430, 2850, 3600 and 1920°C. Match each oxide of Mg, Ca, Sr and Ba with its correct melting point and explain the melting point order. **[4]**

13.4 On heating in excess oxygen, barium forms a peroxide, BaO_2, but magnesium forms only a simple oxide, MgO.
(a) What is the formula of the peroxide ion? **[1]**
(b) Suggest an equation for the reaction between barium peroxide and water. **[2]**
(c) Explain the reluctance of magnesium to form a peroxide. **[4]**

13.5 **(a)** Write an equation for the thermal decomposition of a Group 2 nitrate, $M(NO_3)_2$ where M represents a Group 2 element. **[3]**
(b) For which metal in Group 2 does this reaction occur **(i)** least easily **(ii)** most easily? **[2]**

13.6   State and explain the change in the solubility of the sulphates of Group 2 with increasing atomic number. **[4]**

13.7   Explain why magnesium carbonate decomposes more readily on heating than strontium carbonate. **[2]**

13.8   This question refers to the elements of Groups 1 and 2 of the Periodic Table.
   (a) Lithium nitrate decomposes on heating to give the same products as are obtained from a Group 2 nitrate.
      (i)   Write a balanced equation for the decomposition of lithium nitrate. **[3]**
      (ii) Suggest one other chemical difference between a compound of this metal and those of the rest of Group 1. **[1]**
   (b) What are the flame colours caused by compounds of calcium and barium? **[2]**
   (c) Suggest explanations for the following.
      (i)   The atomic radii of lithium and magnesium are very similar. **[3]**
      (ii) Caesium reacts more vigorously than sodium with water. **[3]**
   (d) Lithium hydride reacts with water according to the equation:

$$LiH(s) + H_2O(l) \rightarrow LiOH(aq) + H_2(g)$$

   1.00 g of impure lithium hydride gave 2.80 dm³ of hydrogen on complete reaction with water. Calculate the percentage purity of the lithium hydride to 3 significant figures.
   (Molar volume of hydrogen = 24.0 dm³ under the conditions of the experiment. Molar mass of lithium = 6.94.) **[4]**

In this section you will need to understand:

- the molecular structures of the elements in Group 7
- the behaviour of the elements as oxidizing agents
- the displacement of some Group 7 elements from their aqueous solutions
- the properties of the hydrogen halides
- how the elements react with water and with aqueous alkali
- the disproportionation of chlorine in aqueous solution.

13.9   Chlorine is a gas at room temperature, while iodine is a solid under the same conditions. Explain the cause of the difference in their physical states. **[3]**

13.10 **(a)** What is meant by the term 'oxidizing agent'? **[2]**
   (b) Write an equation which shows chlorine oxidizing iodide ions. **[2]**
   (c) Why do the halogens become less powerful oxidizing agents with increasing atomic number? **[3]**
   (d) **(i)**   Which halide ion is the most powerful reducing agent? **[1]**
      (ii) Write a balanced equation for the reaction between this halide ion and bromine. **[2]**

13.11 **(a)** Hydrogen chloride can be prepared in the laboratory from sodium chloride.

        (i) Name the other reagent required and state the reaction conditions. **[2]**

        (ii) Write a balanced equation for the reaction in which hydrogen chloride is formed. **[2]**

    (b) In a similar reaction involving sodium bromide, red-brown fumes and sulphur dioxide are formed, and the yield of hydrogen bromide is poor.

        (i) Write half equations for the formation of the red-brown fumes and the sulphur dioxide and hence write a balanced equation for the redox reaction which occurs. **[6]**

        (ii) Explain the differing behaviour of sodium chloride and sodium bromide in these reactions. **[3]**

13.12 Hydrogen fluoride dissolves in water to form an acidic solution but the resulting solution is more weakly acidic than that formed when hydrogen chloride dissolves in water. Explain this observation. **[3]**

13.13 Write a balanced equation for the reaction between chlorine and aqueous sodium hydroxide solution at room temperature. Use the concept of oxidation states to show that a disproportionation reaction has taken place. **[5]**

14 The trends in the physical properties of the elements across a period

- the classification of elements as atomic, simple molecular or giant covalent structures
- how the structure of an element affects its physical properties
- the trends in ionization energy across a period.

14.1 **(a)** Classify the following elements as having either atomic (A), metallic (M), simple molecular (SM) or giant molecular structures (GM): lithium, carbon, nitrogen, oxygen, fluorine and neon. **[6]**

(b) Account for the difference in melting point between carbon (3730 °C) and nitrogen (−210 °C) in terms of their structures. **[4]**

14.2 **(a)** Sodium, magnesium, aluminium and silicon are successive elements in period 3. The table shows their relative electrical conductivities. Comment on these values in terms of the atomic configurations and structures of the elements.

element	relative electrical conductivity
sodium	1.00
magnesium	1.23
aluminium	1.96
silicon	0.50

[4]

(b) Graphite is a soft electrical conductor, while the other carbon allotrope, diamond, is hard and an electrical insulator. Account for these facts. **[6]**

14.3 From the graph of first ionization energy against atomic number explain:

(a) the general trend in first ionization energy from Li to Ne **[3]**

(b) the two irregularities in the graph. **[6]**

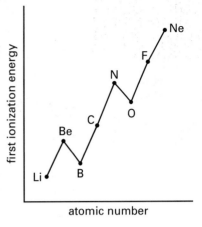

14.4 Below is a list of the metallic or covalent radii and boiling points of the elements Na to Cl of period 3, but not in the correct order:

atomic radius/nm 0.191; 0.143; 0.160; 0.110; 0.117; 0.104; 0.099

boiling point/K 1380; 2630; 1163; 238; 718; 553; 2740.

Copy and complete the table by inserting the correct values for each element. **[4]**

	Na	Mg	Al	Si	P	S	Cl
atomic radius/nm							
boiling point/K							

14.5 The successive ionization energies, in kJ/mol, for four elements are listed below. For each element give its outer electronic configuration and the group in the Periodic Table to which it belongs.

	element			
ionization energy	1	2	3	4
1	519	736	1000	1090
2	7300	1450	2260	2350
3	11 800	7740	3390	4610
4		10 500	4540	6220
5		13 600	6990	37 800
6			8490	47 000
7			27 100	
8			31 700	**[8]**

14.6 The boiling points of the elements of the second short period, lithium to neon are as follows:

element	Li	Be	B	C	N	O	F	Ne
boiling point/°C	1330	2480	3900	4830	−196	−183	−188	−246

(a) Represent these boiling points graphically and comment on the shape of the resulting plot. [6]

(b) (i) Account for the trend in boiling point from beryllium to carbon. [2]

(ii) What type of structure for these elements is suggested by their boiling points? Explain your answer. [3]

(c) Explain the difference in boiling point between carbon and nitrogen. [3]

14.7 The first three ionization energies (in kJ mol^{-1}) of magnesium and aluminium are given below:

	1st	2nd	3rd
Mg	736	1450	7740
Al	577	1820	2740

Explain the following:

(a) the lower first ionization energy of aluminium [3]

(b) the higher second ionization energy of aluminium [3]

(c) the lower third ionization energy of aluminium. [2]

14.8 The melting and boiling points of the elements of the third short period, sodium to argon are as follows:

element	Na	Mg	Al	Si	P	S	Cl	Ar
melting point/°C	98	650	660	1410	44	119	−101	−189
boiling point/°C	892	1110	2450	2680	280	445	−35	−186

(a) Suggest explanations for the following:

(i) the increase in boiling point from sodium to aluminium [3]

(ii) the similarities between the melting points of magnesium and aluminium compared to the considerable difference in their boiling points [5]

(iii) the closeness of the melting and boiling points of argon. [1]

(b) Would you expect trends in boiling points of the elements in a period to be more or less reliable than the trends in melting points as a guide to the relative magnitudes of the interatomic or intermolecular forces present? Explain your answer. [3]

15 Nomenclature and formulae of alkanes and alkenes

In this section you will need to understand:

- how to name and write formulae for straight and branched chain alkanes
- how to name and write formulae for straight and branched chain alkenes
- how to name and write formulae for structural isomers of alkanes and alkenes.

15.1 **(a)** Write the names for the compounds represented by the following structural formulae:

    (i) $CH_3CH_2CH_2CH_2CH_3$ **[1]**

    (ii) $CH_3CH_2CH_2CHCH_3$ **[1]**
                           |
                         CH_3

    (iii) $CH_3CH_2CHCH_2CH_3$ **[1]**
                        |
                     CH_3

        (iv) $CH_3CHCH_2CH_2CH_3$ **[1]**
             |
      CH_3

(b) Which **two** of the above formulae represent the same compound?   **[1]**

(c) Which of the above compounds does *not* have the same empirical (simplest) formula as compound **(iv)**? Write its empirical formula.   **[1]**

(d) Compounds **(ii)** and **(iii)** have the same number of carbons and hydrogens, but are different compounds. They are called isomers and have the same molecular formula. Write the structural formulae of the **two** other isomers for compound **(i)** and give their names.   **[4]**

15.2 **(a)** Write the structural formula for:

    (i) hex-1-ene   **[1]**

    (ii) hex-2-ene   **[1]**

    (iii) hex-3-ene   **[1]**

    (iv) 3-methylhex-2-ene   **[1]**

    (v) 4-methylhex-1-ene.   **[1]**

(b) Write the names of the following alkenes:

    (i) $CH_3CH{=}CHCH_3$ **[1]**

    (ii) $CH_2{=}CHCH_2CH_3$ **[1]**

    (iii) $CH_2{=}CCH_2CH_3$ **[1]**
                    |
               CH_3

    (iv) $CH_3CH_2CH{=}CHCH_3$ **[1]**

(c) Explain why it is not correct to name an alkene but-3-ene. [2]

(d) A straight chain alkene with five carbon atoms is a pentene. Write the formulae for the:

    (i) two possible straight chain five carbon alkenes [2]

    (ii) two branched chain alkenes containing five carbons. [2]

15.3 The C_8 straight chain hydrocarbon is octane. There are however three heptanes, six hexanes, five pentanes and one butane containing eight carbon atoms. Draw up a table showing the structural formulae and name of each one. [15]

15.4 Write the structural formulae of the five isomers of pentene, and name each one. [5]

15.5 **(a)** Write the structural formulae for

    (i) butane [1]

    (ii) 2-methylpropane. [1]

  (b) **(i)** What is the molecular formula for butane and 2-methylpropane? [1]

    (ii) Why are butane and 2-methylpropane described as isomers of each other? [1]

15.6 Write the names of the isomers of C_4H_8. [3]

16

Nomenclature and formulae of open chain compounds containing functional groups

In this section you will need to understand:

For AS
- how to name and write formulae for straight and branched chain alcohols, using suffixes and prefixes
- how to name and write formulae for straight and branched chain halogenoalkanes

For A2
- how to name and write formulae for straight and branched chain amines
- how to name and write formulae for straight and branched chain carbonyl compounds; ketones, aldehydes, carboxylic acids, esters and amides.

16.1 **(a)** Write the names for the following structural formulae:

(i) $CH_3CH_2CH_2OH$

(ii) 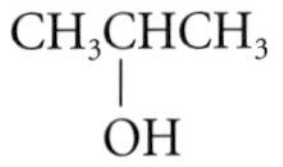 CH_3CHCH_3
         |
         OH

(iii) 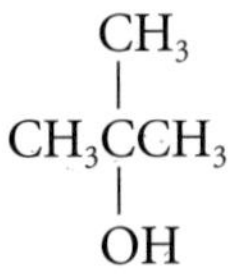
$$CH_3$$
      |
CH_3CCH_3
      |
     OH

(iv) 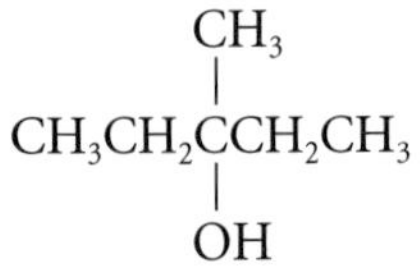
$$CH_3$$
      |
$CH_3CH_2CCH_2CH_3$
      |
     OH

(v)  $CH_3CHCH_2CH_2CH_3$
         |
         OH [5 × [1]]

(b) Write the structural formulae for:

(i) ethanol   **(ii)** propan-1-ol   **(iii)** propan-2-ol

(iv) 1-hydroxybutane   **(v)** 2-hydroxybutane.   [5 × [1]]

(c) Explain why there is no need for the name 3-hydroxybutane or butan-3-ol.   [1]

16.2 Antifreeze contains glycol, which was its old chemical name. The formula of glycol is $HOCH_2CH_2OH$. There are now two ways to name 'glycol' showing the presence of alcohol groups by using either a suffix or a prefix; ethane-1,2-diol or 1,2-dihydroxyethane. Its systematic and preferred name is ethane-1,2-diol.

Write both the suffix name and the prefix name for each of the following:

(a) $HOCH_2CH_2CH_2OH$

(b) 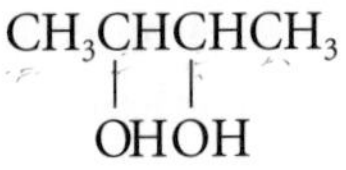 $CH_3CHCHCH_3$
         | |
         $OHOH$

(c) $CH_3CHCH_2CH_2OH$
$\quad\quad\;|$
$\quad\quad OH$

$\quad\quad\quad\quad\quad\quad\quad\quad\quad CH_3$
$\quad\quad\quad\quad\quad\quad\quad\quad\quad\;|$
(d) CH_3CCH_2OH
$\quad\quad\quad\quad\quad\quad\quad\quad\quad\;|$
$\quad\quad\quad\quad\quad\quad\quad\quad\quad OH$ **[4 × [2]]**

(e) Having tried to write both names for each formula suggest which is the better method of naming diols. **[1]**

16.3 Name the following chlorinated hydrocarbons:
(a) $CH_3CH_2CH_2CH_2Cl$

(b) $CH_3CHCH_2CH_3$
$\quad\quad\;|$
$\quad\quad Cl$

(c) $ClCH_2CH_2CH_3$

(d) $CH_3CHCH_2CHCH_3$
$\quad\quad\;\;|\quad\quad\;|$
$\quad\quad\;Cl\quad\quad Cl$

$\quad\quad CH_3$
$\quad\quad\;|$
(e) CH_3CCH_2Cl
$\quad\quad\;|$
$\quad\quad Cl$ **[5 × [1]]**

16.4 **(a)** Name the following dibromoalkanes:
(i) $BrCH_2CH_2CH_2CH_2Br$

(ii) $CH_3CHCH_2CH_2Br$
$\quad\quad\quad\;|$
$\quad\quad\quad Br$

(iii) $Br\,CHCH_2CH_3$
$\quad\quad\;\;|$
$\quad\quad Br$

(iv) $CH_3CHCH_2CHCH_3$
$\quad\quad\quad\;|\quad\quad\;|$
$\quad\quad\;\;Br\quad\;\;Br$

$\quad\quad CH_2Br$
$\quad\quad\;|$
(v) CH_3CCH_3
$\quad\quad\;|$
$\quad\quad Br$ **[5 × [1]]**

(b) Which ones of **(i)–(v)** are isomers of each other? **[1]**

16.5 Write the structural formulae for:
(a) butanal
(b) butanone
(c) pentanal
(d) pentan-2-one and pentan-3-one
(e) butanoic acid
(f) 2-hydroxybutanoic acid, 3-hydroxybutanoic acid and 4-hydroxybutanoic acid. **[9]**

16.6 **(a)** Write the structural formulae for:
(i) ethyl ethanoate
(ii) methyl ethanoate
(iii) methyl methanoate
(iv) ethyl methanoate. **[4 × [1]]**
(b) Which compounds are isomers of each other? **[1]**

16.7 Write the structural formulae for:
(a) aminoethane
(b) 2-aminopropane
(c) ethanamide
(d) propanamide. **[4 × [1]]**

17 Reactions of alkanes and alkenes and bonding between carbon atoms

In this section you will need to understand:

- addition reactions of alkenes
- the oxidation of alkanes and alkenes
- the effects of substituents on the manner of addition to alkenes
- bonding in single, double and triple bonds
- the effects of substituents on the electron density of bonding in functional groups
- free radical reactions of alkanes
- bond energy calculations and heats of combustion for the reactions of alkanes and alkenes
- sp hybrid orbitals and their shapes
- delocalization effects in open chain compounds, dienes and delocalization of bonds.

17.1 Complete the following equations:

(a) $CH_2{=}CH_2 + Cl_2 \rightarrow$ [1]

(b) $CH_3CH{=}CH_2 + Cl_2 \rightarrow$ [1]

(c) $CH_2{=}CH_2 + HCl \rightarrow$ [1]

(d) $CH_3CH{=}CH_2 + HCl \rightarrow$ [2]

(e) $CH_2{=}CHCH{=}CH_2 + 2Cl_2 \rightarrow$ [1]

17.2* By showing the mechanism of addition suggest an explanation for each of the following statements:

(a) the addition of HCl to propene forms, as the major product, 2-chloropropane rather than 1-chloropropane [5]

(b) the addition of 1 mol of Br_2 to buta-1,3-diene forms 1,4-dibromobut-2-ene rather than 1,2-dibromobuta-3-ene. [3]

17.3 Write the structural formulae, including approximate bond angles, for each of the molecules (a) to (e), following the example given for methane.

methane

$$\begin{array}{c} H \\ {}_{109}\ |\ {}_{109} \\ H - C - H \\ {}_{109}\ |\ {}_{109} \\ H \end{array}$$

(a) ethane [1]

(b) ethene [1]

(c) ethyne [1]

(d) chloromethane [1]

(e) chloroethane [1]

17.4* Draw the shapes of the following orbitals:

(a) (i) s [1]

(ii) p_x p_y p_z [3]

(b) (i) sp^3 hybrids [1]

(ii) sp^2 hybrids [1]

(iii) sp hybrids. [1]

(c) Which of the orbitals stated in parts (a) and (b) are around a carbon atom in:

(i) methane [1]

(ii) octane [1]

(iii) ethene [1]

(iv) ethyne? [1]

17.5 Describe or name the shape of each of the following molecules:

(a) methane [1]

(b) ethane [1]

(c) ethene [1]

(d) tetrachloromethane [1]

(e) dimethylpropane. [1]

17.6 (a) What are the names of the four products which can be formed when a mixture of bromine vapour and methane is exposed to UV light? [4]

(b) What is the name of the type of mechanism for the reaction? [1]

(c) Write the five steps in the three-stage mechanism for the formation of chloromethane from a mixture of chlorine and methane which is exposed to UV light:

(i) initiation [1]

(ii) the two propagation steps [2]

(iii) the two possible termination steps. [2]

17.7 Write balanced equations for the complete combustion of:

(a) methane [1]

(b) propane [1]

(c) octane [1]

(d) propene [1]

(e) penta-1,3-diene. [1]

17.8 The enthalpy changes of combustion (heats of combustion) for a number of alkanes in the homologous series of alkanes are given in this table:

straight chain alkane	enthalpy changes of combustion $kJ\ mol^{-1}$
methane	-890
ethane	-1560
propane	-2219
butane	-2877
pentane	-3509
hexane	-4195
heptane	
octane	-5512
nonane	-6125
decane	-6778

(a) Plot the enthalpy change of combustion against the number of carbon atoms in each molecule and comment on the shape of the graph obtained. **[4]**

(b) The enthalpy change of combustion for heptane has been omitted from the table. Use your graph to determine a value for heptane. **[1]**

(c) Petrol can be considered as a mixture of isomers containing eight carbon atoms. Details of the enthalpy changes of combustion of three branched chain alkanes containing four or five carbon atoms per molecule are given in the table below. Suggest which isomer containing eight carbon atoms is likely to have the largest enthalpy change of combustion value, and write its structural formula. **[2]**

branched chain alkane	enthalpy change of combustion $kJ\ mol^{-1}$
2-methylpropane	-2868
2-methylbutane	-3503
2,2-dimethylpropane	-3517

17.9

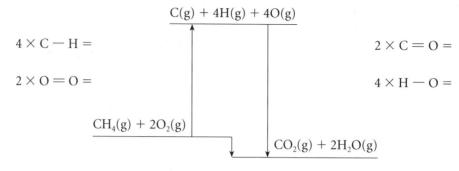

$C(g) + 4H(g) + 4O(g)$

$4 \times C - H =$

$2 \times O = O =$

$2 \times C = O =$

$4 \times H - O =$

$CH_4(g) + 2O_2(g)$

$CO_2(g) + 2H_2O(g)$

bond	average bond energy kJ mol^{-1}
C—C	346
C—H	413
O=O	497
C=O	803
O—H	464

(a) Using the figures in the table above, copy and complete the cycle to determine an enthalpy charge of combustion value for methane. **[3]**

(b) Construct and use an energy cycle to find the enthalpy change of combustion of:

(i) propane **[6]**

(ii) octane. **[6]**

In each case comment on the value you obtain compared with the values quoted in the table in question **17.8**.

18 Reactions of alcohols

In this section you will need to understand:

- substitution reactions of straight and branched chain alcohols
- elimination reactions of straight and branched chain alcohols
- oxidation reactions of straight and branched chain alcohols.

18.1 What are the names of the compounds formed by substitution reactions, using PCl_5, with each of the following alcohols:

(a) ethanol [1]

(b) propan-1-ol [1]

(c) propan-2-ol [1]

(d) each of the three isomers of straight chain hexanols [3]

(e) each of the eight $C_5H_{11}OH$ isomers? [8]

18.2 (a) Complete the following equations:

(i) $CH_3CH_2CH_2OH + HI \rightarrow$ [1]

(ii) $CH_3CH_2CH_2OH + HBr \rightarrow$ [1]

(iii) $CH_3CHCH_3 + HBr \rightarrow$ [1]
$\qquad$ |
$\qquad$ OH

(b) The reactions shown in parts (ii) and (iii) above are reversible reactions. Explain why the use of sodium bromide and concentrated sulphuric acid, rather than of HBr itself, helps the forward reaction. [2]

18.3 Complete the following equations to show the organic product(s) in each case.

(a) $CH_3CHCH_3 + P_2O_5 \rightarrow$ [1]
$\qquad$ |
$\qquad$ OH

(b) $CH_3CHCH_2CH_3 + P_2O_5 \rightarrow$ [2]
$\qquad$ |
$\qquad$ OH

(c) $CH_3CH_2CH_2CH_2OH + P_2O_5 \rightarrow$ [1]

18.4* Alcohols undergo nucleophilic substitution. Useful reagents to use are:
phosphorus pentachloride,
phosphorous trichloride,
sodium bromide and concentrated sulphuric acid,
a mixture of red phosphorus and iodine,
hydriodic acid.

(a) Write the formulae for the three nucleophiles present in the reagents
above. [1]

(b) (i) Write the S_N2 mechanism for the substitution of a primary alcohol, such
as ethanol, by HBr. [2]

(ii) Write the S_N1 mechanism for the substitution of a tertiary alcohol, such
as 2-hydroxy-2-methylpropane, by HBr. [2]

(iii) Explain why tertiary alcohols are more likely to go by the S_N1
mechanism than the S_N2 mechanism. [2]

(iv) Why do reactions which go by the S_N1 mechanism usually lead to a
poorer yield of halogenoalkane than the S_N2 mechanism? [1]

18.5* **(a)** Copy and complete the table below to give the name and formula of each
product formed (if any), when the alcohol is heated under reflux with acidic
dichromate.

alcohol	structural formula of alcohol	name of product	formula of product	
methanol				[3]
ethanol				[3]
propan-1-ol				[3]
propan-2-ol				[3]
2-methylpropan-2-ol				[2]

(b) For the five alcohols above complete a second table which shows the
products (if any), when the reaction conditions are altered such that the
product is distilled out as it is formed. [8]

19 Reactions of halogenoalkanes and halogenoalkenes

In this section you will need to understand:

- the reactions of straight and branched chain halogenoalkanes
- the reactions of straight and branched chain halogenoalkenes.

19.1 Describe the reactions of OH^- with CH_3CH_2CHBr to illustrate:

$$|$$
$$CH_3$$

    (i) the S_N1 mechanism   **[4]**

    (ii) the S_N2 mechanism.   **[4]**

    (You will find it helpful to use 'curly arrows' in your answer.)

19.2 Chloroalkanes undergo either substitution reactions or elimination reactions when they are treated with a base.

    (a) Complete the equations below to show all the organic products which may be formed.

        (i) $CH_3CH_2CH_2Cl + NaOH \rightarrow$ **[2]**

        (ii) $CH_3CHCH_3 + NaOH \rightarrow$ **[2]**

              $|$

              Cl

        (iii) $CH_3CHCH_2CH_3 + NaOH \rightarrow$ **[3]**

                $|$

               Cl

    (b) State how the conditions can be altered to favour substitution rather than elimination.   **[2]**

    (c) State why the reaction below strongly favours the elimination product rather than the substitution product.

$$CH_3$$
$$|$$
$$CH_3CCH_3 + NaOH \rightarrow$$
$$|$$
$$Cl$$

                                        [1]

19.3* The compound vinyl chloride is the monomer for making polyvinylchloride, pvc. One industrial process for making vinyl chloride can be described by the following equations:

$$CH_2{=}CH_2 + Cl_2 \xrightarrow{\text{step 1}} ClCH_2CH_2Cl \xrightarrow{\text{step 2}} CH_2{=}CHCl$$

The second step is done by heating the 1,2-dichloroethane at 600 °C and passing it over pumice.

(a) What is the systematic name for vinyl chloride? [1]

(b) In the industrial process, what type of reaction is:

   (i) the first step [1]

   (ii) the second step? [1]

(c) What would be formed if two moles of HCl had been eliminated from the 1,2-dichloroethane instead of one? [1]

(d) In small scale or laboratory work, the normal reagent for step 2 is sodium hydroxide dissolved in ethanol. If this were used, what substitution products could be expected? [2]

19.4 An investigation was set up to see which reacted most quickly with dilute alkali; 1-bromobutane, 1-chlorobutane or 1-iodobutane.

0.01 mol of each halogenoalkane was added to 20 cm^3 of dilute sodium hydroxide solution and the mixtures shaken for 30 seconds. A 1 cm^3 portion was then removed, added to dilute nitric acid and silver nitrate solution and the amount of precipitate formed observed.

(a) Which two organic products are formed in each case? [2]

(b) What colour are the precipitates formed in each case? [3]

(c) Which reaction will go fastest? Explain why. [3]

(d) (i) What is the mass of 0.01 mol of 1-bromobutane? [1]

   (ii) What mass of sodium bromide would be expected if the reaction went to completion? [1]

Part 2 A2 Questions

20 Energetics 2

> ### In this chapter you will need to understand:
>
> - what is meant by the term 'lattice enthalpy'
> - what is meant by the terms 'ionization energy' and 'electron affinity' and to have a working knowledge of the various enthalpy terms previously met at AS level
> - what is meant by a Born–Haber cycle and how such a cycle can be constructed for an ionic crystalline lattice in order to determine its lattice enthalpy
> - how lattice enthalpies vary with the size and charge of the ions contained in the lattice
> - how to use a Born–Haber cycle to determine enthalpy changes of formation for possible compounds and to use the results to compare their likely energetic stabilities
> - how a comparison of lattice enthalpies calculated for a specific ionic lattice by means of a Born–Haber cycle and by a theoretical method can provide information on the extent to which the ions in the lattice deviate from a spherical shape and to be able to suggest factors which may be responsible for this.

20.1 What is meant by the term 'lattice enthalpy'? **[3]**

20.2 Lattice enthalpies can be calculated using a Born–Haber cycle. Such a cycle requires various information to be put into it, including ionization energy, electron affinity and standard enthalpy change of atomization. State what is meant by the following terms, illustrating each of your answers with a suitable equation to represent the change you have described:

(a) first ionization energy **[4]**

(b) second ionization energy **[4]**

(c) first electron affinity **[4]**

(d) standard enthalpy change of atomization. **[4]**

20.3 The first electron affinity value for the formation of a singly charged sulphide ion, S^-, from elemental sulphur is $-200.4 \text{ kJ mol}^{-1}$ but the second electron affinity value for the formation of a doubly charged sulphide ion, S^{2-}, is $+640 \text{ kJ mol}^{-1}$. Explain the difference in values. **[3]**

20.4 A simple Born–Haber cycle is of the form:

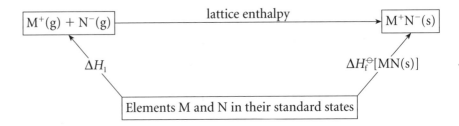

(a) State the law upon which the use of a Born–Haber cycle depends. **[3]**

(b) Using the law which you have just stated, write down a relationship between the quantities: lattice enthalpy, ΔH_1, and $\Delta H_f^\ominus$ [MN(s)]. **[1]**

(c) Write down the various energy quantities which you would need in order to determine the value of ΔH_1 before using ΔH_1 in the energy cycle. **[4]**

20.5 A Born–Haber cycle for potassium bromide is shown below:

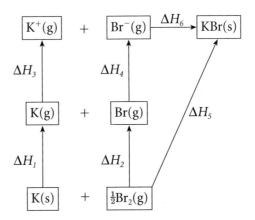

(a) Name each energy quantity ΔH_1 to ΔH_5 inclusive. **[5]**

(b) The values of the energy quantities ΔH_1 to ΔH_5 are shown below;

$\Delta H_1 = +89.2$ kJ mol^{-1}, $\Delta H_2 = +111.9$ kJ mol^{-1},

$\Delta H_3 = +419$ kJ mol^{-1}, $\Delta H_4 = -324.6$ kJ mol^{-1},

$\Delta H_5 = -393.8$ kJ mol^{-1}.

Use these values to calculate ΔH_6. **[2]**

Data: The data required for the following questions are included in the tables below (all in kJ mol^{-1}):

E_{m1}	E_{m2}	E_{m3}	$\Delta H_{at}^{\ominus}$
Al +578	Al$^+$ +1817	Al^{2+} +2745	Al +326.4
Na +496	Na$^+$ +4563	Na^{2+} +6913	Na +107.3
Rb +403	Rb$^+$ +2632	Rb^{2+} +3900	Rb +80.9
Sr +550	Sr$^+$ +1064	Sr^{2+} +4210	Sr +164.4
Zn +906	Zn$^+$ +1733	Zn^{2+} +3833	Zn +130.7

E_{aff1}	E_{aff2}	$\Delta H_f^{\ominus}$	$\Delta H_{at}^{\ominus}$
F −328.0		Al$_2$O$_3$(s) = −1675.7	F +79.0
I −295.4		Na$_2$O(s) = −414.2	I +106.8
O −141.1	O$^-$ +798	RbF(s) = −557.7	O +249.2
S −200.4	S$^-$ +640	SrS(s) = −453.1	S +278.8
		ZnI$_2$(s) = −208.0	

20.6 Calculate the lattice enthalpy of:
   (a) rubidium fluoride, RbF   **[4]**
   (b) sodium oxide, Na$_2$O   **[4]**
   (c) zinc iodide, ZnI$_2$ **[4]**
   (d) strontium sulphide, SrS   **[4]**
   (e) aluminium oxide, Al$_2$O$_3$.   **[4]**
In each case give your answer to the nearest whole number.

20.7 Look at the lattice enthalpy values given in the tables below (all in kJ mol^{-1}):

LiF −1036	LiCl −853	LiBr −807
NaF −923	NaCl −786	NaBr −747
KF −821	KCl −715	KBr −682

Li$_2$O −2814	BeO −4443	BeCl$_2$ −3020
Na$_2$O −2478	MgO −3791	MgCl$_2$ −2526
K$_2$O −2232	CaO −3401	CaCl$_2$ −2258

   (a) Identify trends in lattice enthalpy values as:
      (i) the size of the cation increases   **[1]**
      (ii) the size of the anion increases   **[1]**
      (iii) the charge on the cation increases   **[1]**
      (iv) the charge on the anion increases.   **[1]**
   (b) Try to explain the trends you have observed.   **[6]**

20.8 CaI_2 is the formula for an energetically stable calcium compound; CaI and CaI_3 are not energetically stable. The latter substances might be expected to have lattice enthalpies similar in value to KI ($-649\,kJ\,mol^{-1}$) and GaI_3 ($-5476\,kJ\,mol^{-1}$) because K, Ca and Ga are all in the same period in the Periodic Table.

(a) By assuming these values, together with a figure of $-2074\,kJ\,mol^{-1}$ for the lattice enthalpy of CaI_2, use Born–Haber cycles to calculate the enthalpy change of formation for each of the three compounds CaI, CaI_2 and $CaCl_3$ correct to the nearest whole number. **[12]**

(b) Comment on the values you obtain in the light of their energetic stabilities. **[9]**

Other data which you will require in the calculation are:

First ionization energy of calcium	=	$+\quad590\,kJ\,mol^{-1}$
Second ionization energy of calcium	=	$+1145\,kJ\,mol^{-1}$
Third ionization energy of calcium	=	$+4912\,kJ\,mol^{-1}$
Electron affinity of iodine	=	$-295.4\,kJ\,mol^{-1}$
Enthalpy change of atomization of calcium	=	$+178.2\,kJ\,mol^{-1}$
Enthalpy change of atomization of iodine	=	$+106.8\,kJ\,mol^{-1}$

20.9 (a) Calculate, by means of a Born–Haber cycle, the enthalpy changes of formation of $CuBr$, CuI, $CuBr_2$ and CuI_2 to the nearest whole number, assuming the following data:

Lattice enthalpy of copper(I) bromide, $CuBr$	=	$-979\,kJ\,mol^{-1}$
Lattice enthalpy of copper(I) iodide, CuI	=	$-966\,kJ\,mol^{-1}$
Lattice enthalpy of copper (II) bromide, $CuBr_2$	=	$-2763\,kJ\,mol^{-1}$
Lattice enthalpy of copper (II) iodide, CuI_2	=	$-2672\,kJ\,mol^{-1}$
First ionization energy of copper	=	$+746\,kJ\,mol^{-1}$
Second ionization energy of copper	=	$+1958\,kJ\,mol^{-1}$
Electron affinity of bromine	=	$-324.6\,kJ\,mol^{-1}$
Electron affinity of iodine	=	$-295.4\,kJ\,mol^{-1}$
Enthalpy change of atomization of copper	=	$+338.3\,kJ\,mol^{-1}$
Enthalpy change of atomization of bromine	=	$+111.9\,kJ\,mol^{-1}$
Enthalpy change of atomization of iodine	=	$+106.8\,kJ\,mol^{-1}$ **[4 × 4]**

(b) (i) In the light of your answers, suggest why when a solution of iodide ions is added to a solution of copper(II) sulphate, copper(I) iodide and iodine are formed instead of copper(II) iodide. **[4]**

(ii) Would you expect a similar result to that described in **(b) (i)** (i.e. the formation of copper(I) bromide and bromine) when bromide ions are added to a solution of copper(II) sulphate? Justify your answer. **[4]**

20.10 Apart from calculating lattice enthalpies by means of a Born–Haber cycle, it is possible to calculate values on the basis of the application of the principles of electrostatics to the gradual build-up of a lattice from individual ions, assuming that the ions are themselves completely spherical.

The following table is a comparison of some values calculated by each of the methods:

substance	lattice enthalpy/kJ mol^{-1}	
	by Born–Haber cycle	by electrostatic method
LiF	−1036	−1030
LiI	−757	−730
NaF	−923	−910
NaI	−704	−682
KF	−821	−808
KI	−649	−632
AgF	−967	−953
AgI	−889	−808
BeF$_2$	−3505	−3150
BeI$_2$	−2800	−2653
MgF$_2$	−2957	−2913
MgI$_2$	−2327	−1944
CaF$_2$	−2630	−2609
CaI$_2$	−2074	−1905
ZnF$_2$	−3032	−2930
ZnI$_2$	−2605	−2549

(a) In each case, calculate the difference in values as a percentage of the value obtained by the electrostatic method. **[4]**

(b) In which cases would a model based on spherical ions appear to give good agreement with the Born–Haber method for obtaining lattice enthalpies? (The values used in the Born–Haber cycle are generally experimentally determined whereas the electrostatic method involves a theoretical calculation.) **[2]**

(c) In which cases does the agreement appear less satisfactory? Suggest why this may be so. (*Hint*: Consider possible reasons for a distortion from a model based on wholly spherical ions.) **[5]**

21 Kinetics 2

In this section you will need to understand:

- how to choose a suitable method for measuring the rate of a chemical reaction
- what is meant by a rate equation and each of the quantities which appear in it
- what is meant by the 'mechanism' of a reaction
- how to determine the 'order' of a reaction with regard to a particular species from given experimental data
- how to determine the overall order for a reaction
- how to decide whether experimentally determined information concerning the order of a reaction confirms or rejects a suggested mechanism for the reaction
- the mechanisms for nucleophilic substitution reactions designated S_N1 and S_N2
- how to determine a first-order rate constant.

21.1 The rate of a reaction can be determined using a variety of different practical techniques. Describe briefly, in each case, a method which would be suitable for monitoring the rate of the following reactions.

(a) The thermal decomposition of potassium chlorate(V),

$$2KClO_3(s) \rightarrow 2KCl(s) + 3O_2(g)$$ [2]

(b) The acid-catalysed hydrolysis of 1,1-diethoxyethane in excess water,

$$CH_3CH(OC_2H_5)_2(l) + H_2O(l) \rightarrow CH_3CHO(l) + 2C_2H_5OH(l)$$ [3]

(c) The thermal decomposition of sulphur dichloride oxide,

$$SO_2Cl_2(g) \rightarrow SO_2(g) + Cl_2(g)$$ [2]

(d) The conversion of α-glucose into an equilibrium mixture of α-glucose and β-glucose. This reaction is carried out in aqueous solution and is catalysed by acids or bases. Both forms of glucose are optically active, they both rotate the plane of polarized light in a clockwise direction as viewed by an observer but, using solutions of the same concentration, α-glucose has a much greater degree of rotation. [2]

(e) The acid-catalysed hydrolysis of methyl ethanoate,

$$CH_3CO_2CH_3(l) + H_2O(l) \rightarrow CH_3CO_2H(l) + CH_3OH(l)$$ [5]

(f) The decomposition of benzenediazonium chloride in aqueous solution,

$$C_6H_5N_2Cl(aq) + H_2O(l) \rightarrow C_6H_5OH(aq) + HCl(aq) + N_2(g)$$ [2]

(g) The reaction between bromide ions and concentrated nitric acid,

$$2Br^-(aq) + 4HNO_3(aq) \rightarrow Br_2(aq) + 2NO_2(g) + 2H_2O(l) + 2NO_3^-(aq)$$ [5]

21.2 A rate equation is of the form:

$$Rate = k[A]^m[B]^n$$

(a) Explain what each of the following symbols means:

    (i) k [2]

    (ii) $[A]$ and $[B]$ [2]

    (iii) m and n. [2]

(b) A rate equation cannot be derived from the stoichiometric equation for a chemical reaction. Why is this so? [2]

(c) How, in principle, would you set about determining the rate equation for a reaction? [4]

21.3 **(a)** What do you understand by the term 'the mechanism of a reaction'? [2]

(b) If a chemical reaction happens to occur in a series of steps, what name is given to the step which has the most significant effect on the overall rate of the reaction? [1]

(c) In what way is a rate equation used to support (or not) a suggested mechanism for a reaction? [2]

21.4 Show that the following data are consistent with the suggested rate equation for the reaction:

$$P + Q \rightarrow R$$

experiment	[P] /mol dm^{-3}	[Q] /mol dm^{-3}	initial rate /mol dm^{-3} s^{-1}
1	0.2	0.2	0.04
2	0.2	0.4	0.08
3	0.4	0.2	0.08
4	0.4	0.4	0.16

Suggested rate equation:

$$Rate = k[P][Q]$$ [3]

21.5 The following table gives the initial rates of reaction for a series of experiments using different concentrations of X and Y for the reaction:

$$X + Y \rightarrow products$$

[X]/mol dm^{-3}	0.15	0.30	0.30	0.45	0.45
[Y]/mol dm^{-3}	0.30	0.30	0.45	0.45	0.75
initial rate/mol dm^{-3} s^{-1}	0.045	0.088	0.090	0.135	0.133

(a) Deduce the rate equation for this reaction and state the overall order. Explain how you reach your conclusions. [4]

(b) Calculate the rate constant for the reaction. State its units clearly showing how you reach your decision. [3]

21.6 What are the overall orders for each of the following reactions?

(a) $2N_2O_5(g) \rightarrow 4NO_2(g) + O_2(g)$
for which the rate equation is:
$$Rate = k[N_2O_5(g)]$$ [1]

(b) $2NO(g) + O_2(g) \rightarrow 2NO_2(g)$
for which the rate equation is:
$$Rate = k[NO(g)]^2[O_2(g)]$$ [1]

(c) $(CH_3)_3CCl(l) + OH^-(aq) \rightarrow (CH_3)_3COH(l) + Cl^-(aq)$
for which the rate equation is:
$$Rate = k[(CH_3)_3CCl(l)][OH^-(aq)]^0$$ [1]

(d) $BrO_3^-(aq) + 5Br^-(aq) + 6H^+(aq) \rightarrow 3Br_2(aq) + 3H_2O(l)$
for which the rate equation is:
$$Rate = k[BrO_3^-(aq)][Br^-(aq)][H^+(aq)]^2$$ [1]

21.7 It is suggested that the decomposition of nitrogen pentoxide dissolved in an organic solvent is a first order reaction. Do the following results, obtained at 45 °C, support this suggestion?

time/s	0	200	400	600	800	
concentration /mol dm^{-3}	2.33	2.04	1.80	1.58	1.40	
time/s	1000	1400	1800	2200	2600	
concentration /mol dm^{-3}	1.25	1.01	0.79	0.59	0.48	[5]

21.8 The oxidation of acidified iodide ions by hydrogen peroxide can be followed by adding a small, but fixed volume, of sodium thiosulphate solution to the reaction mixtures together with 1 cm^3 of starch solution and measuring the time taken for a blue colour to appear.

The equation for the oxidation is:

$$H_2O_2(aq) + 2H^+(aq) + 2I^-(aq) \rightarrow I_2(aq) + 2H_2O(l)$$

In a series of experiments, the following results were obtained:

volume of 1.0 M H$_2$SO$_4$(aq)/cm^3	volume of 0.1 M KI(aq)/cm^3	volume of water/cm^3	volume of 0.1 M H$_2$O$_2$(aq)/cm^3	time taken /s
10.0	5.0	20.0	5.0	77
10.0	10.0	15.0	5.0	48
10.0	15.0	10.0	5.0	31
10.0	20.0	5.0	5.0	22

(a) Why is a blue colour formed in the reaction? **[2]**

(b) Explain why the formation of this blue colour can be used to monitor the rate of the reaction. **[1]**

(c) Why is the volume of water used not kept constant? **[2]**

(d) Using the above data the order of the reaction with respect to iodide ions can be determined but not the order with respect to hydrogen peroxide or hydrogen ions. Explain why this is so. **[1]**

(e) What further experiments would have to be carried out in order to determine the orders of reaction with respect to hydrogen peroxide and to hydrogen ions? **[4]**

(f) Determine the rate of each reaction from the data listed. **[2]**

(g) What conclusion can you reach about the order of the reaction with respect to iodide ions? **[4]**

21.9 The rate equation for the reaction between bromate(V) and bromide ions in acidic solution is:

$$\text{Rate} = k[BrO_3^-][Br^-][H^+]^2$$

Explain whether this is consistent with the suggested mechanism for the reaction which is shown below:

1	$H^+ + Br^- \rightarrow HBr$	fast
2	$H^+ + BrO_3^- \rightarrow HBrO_3$	fast
3	$HBr + HBrO_3 \rightarrow HOBr + HOOBr$	slow
4	$HOOBr + HBr \rightarrow 2HOBr$	fast
5	$HOBr + HBr \rightarrow H_2O + Br_2$	fast

[5]

21.10 Propanone undergoes an acid-catalysed substitution reaction with iodine.

$$CH_3COCH_3(aq) + I_2(aq) \rightarrow CH_3COCH_2I(aq) + HI(aq)$$

(a) Suggest **two** methods that would be suitable for determining the order of this reaction.

(b) Use the following results to determine the order of reaction with respect to propanone, iodine and hydrogen ions. **[3]**

[propanone] /mol dm^{-3}	[iodine] /mol dm^{-3}	[hydrogen ions] /mol dm^{-3}	relative rate of reaction
0.30	0.30	0.15	4
0.30	0.15	0.15	4
0.15	0.30	0.15	2
0.15	0.30	0.075	1

(c) Write the rate equation for the reaction. **[1]**

(d) Suggest a mechanism for this reaction which would be consistent with the information contained in the rate equation. **[5]**

21.11 Nucleophilic substitution reactions can take place by two main mechanisms which are known as S_N1 and S_N2. These mechanisms are illustrated below.

S_N1 $R—X \rightarrow R^+ + X^-$ (slow)

then $R^+ + OH^- \rightarrow ROH$ (fast)

S_N2 $^-HO + R—X \rightarrow [HO--- R --- X] \rightarrow HO—R + X^-$ (slow)

Explain how a rate equation can help to confirm which mechanism is likely to be followed by a particular halogenoalkane. **[4]**

21.12 The reactions of the two isomers, 1-chloro-1-phenylethane and 1-chloro-2-phenylethane, with sodium hydroxide solution have been extensively investigated. A solution of each of the isomers was mixed with sodium hydroxide solution and allowed to stand at a temperature of 30 °C. It was arranged that each solution had a concentration of 0.1 mol dm^{-3} with respect to both the organic compound and the alkali. At known times samples of the mixture were withdrawn, quenched and titrated with a standard solution of an acid.

The results obtained are shown below:

1-chloro-1-phenylethane	
time/hours	titre/cm^3
0.2	21.0
1.0	19.8
3.0	17.3
7.0	13.4
12.0	9.4
18.0	6.2
22.0	4.7

1-chloro-2-phenylethane	
time/hours	titre/cm^3
0.2	23.2
20.0	20.4
50.0	17.6
100.0	14.8
200.0	10.3
350.0	7.0
500.0	5.6

(a) By plotting suitable graphs, determine whether these reactions are zero order, first order or second order. **[8]**

(b) Are your answers overall orders or orders with respect to a single reactant? Explain you answer. **[2]**

(c) Hence determine in each case whether the likely mechanism is S_N1 or S_N2. **[2]**

(d) Determine the value of the rate constant for the reaction using 1-chloro-1-phenylethane. Your answer should include the correct units. **[4]**

Equilibrium 2

In this section you will need to understand:

- the use of the 'equilibrium law' and the term 'equilibrium constant', expressed both in terms of concentrations and of partial pressures
- the application of the equilibrium law to homogeneous and heterogeneous equilibria
- how to calculate equilibrium constants from suitable data and to appreciate the significance of their magnitude in terms of the position of equilibrium in a reaction
- the application of Le Chatelier's principle to suitable reactions undergoing a change of experimental conditions
- the meaning of the terms 'strong acid' and 'weak acid'
- the meaning of the term pH and its calculation for strong acids and strong alkalis
- the meaning of the term K_a for a weak acid and its use in determining the pH of weak acids; also how to carry out the reverse calculation from pH to K_a
- what is meant by the term 'buffer solution' and to be able to carry out simple calculations concerning such solutions.

22.1 **(a)** What is 'the equilibrium law'? **[2]**

(b) Use the equilibrium law to write down the expression for the equilibrium constant, K_c, for the reaction:

$$m\text{A} + n\text{B} \rightleftharpoons p\text{C} + q\text{D}$$ **[2]**

(c) What is the significance of the size of the equilibrium constant? **[2]**

22.2 Write down the expression for the equilibrium constant, K_c, for each of the following reactions:

(a) $CH_3OH(l) + C_2H_5CO_2(l) \rightleftharpoons C_2H_5CO_2CH_3(l) + H_2O(l)$ **[3]**

(b) $C_5H_{10}(l) + CH_3CO_2H(l) \rightleftharpoons CH_3CO_2C_5H_{11}(l)$ **[3]**

(c) $2O_3(g) \rightleftharpoons 3O_2(g)$ **[3]**

(d) $N_2O_4(\text{trichloroethane}) \rightleftharpoons 2NO_2(\text{trichloroethane})$ **[3]**

(e) $Cu^{2+}(aq) + 4NH_3(aq) \rightleftharpoons [Cu(NH_3)_4^{2+}](aq)$ **[3]**

In each case, state in what units K_c would be expressed.

22.3 Write down the expression for the equilibrium constant, K_c, for the following reactions in terms of the partial pressures of each substance present:

(a) (i) $PCl_5(g) \rightleftharpoons PCl_3(g) + Cl_2(g)$ [3]

(ii) $CO(g) + Br_2(g) \rightleftharpoons COBr_2(g)$ [3]

(iii) $4PF_5(g) \rightleftharpoons P_4(g) + 10F_2(g)$ [3]

(iv) $2H_2S(g) \rightleftharpoons 2H_2(g) + S_2(g)$ [3]

(v) $4NH_3(g) + 5O_2(g) \rightleftharpoons 4NO(g) + 6H_2O(l)$ [3]

In each case, state in what units K_p would be expressed if the partial pressures and total pressures are given in atmospheres.

(b) All the reactions given in **(a)** (and also in **22.1**) are examples of 'homogeneous equilibria'. What do you understand by this term? [1]

22.4 The examples which now follow are of 'heterogeneous equilibria'.

(a) What do you understand by the term 'heterogeneous equilibria'? [1]

(b) In each of the following examples, write down the expression for the equilibrium constant, K_c.

(i) $Br_2(g) \rightleftharpoons Br_2(l)$ [2]

(ii) $Zn(s) + Cu^{2+}(aq) \rightleftharpoons Zn^{2+}(aq) + Cu(s)$ [2]

(iii) $Ag_3PO_4(s) \rightleftharpoons 3Ag^+(aq) + PO_4^{3-}(aq)$ [2]

(iv) $Ag^+(aq) + Fe^{2+}(aq) \rightleftharpoons Ag(s) + Fe^{3+}(aq)$ [2]

(c) The heterogeneous equilibria below relate to thermal decomposition reactions. Write down the expression for the equilibrium constant, K_p, for the following reactions:

(i) $CaCO_3(s) \rightleftharpoons CaO(s) + CO_2(g)$ [2]

(ii) $NH_2CO_2NH_4(s) \rightleftharpoons 2NH_3(g) + CO_2(g)$ [2]

22.5 In a classic investigation of the equilibrium involving ethanoic acid, ethanol, ethyl ethanoate and water, Berthelot and St Gilles took known amounts of acid and alcohol and sealed them in glass tubes which were then heated to over 100 °C (over 373 K). The tubes were then rapidly cooled and the contents analysed. The equation for the reaction is:

$$CH_3CO_2H(l) + C_2H_5OH(l) \rightleftharpoons CH_3CO_2C_2H_5(l) + H_2O(l)$$

In one of their experiments, at approximately 400 K, they started with 1.00 mol of ethanoic acid and 0.500 mol of ethanol. At equilibrium, they found that 0.414 mol of ethyl ethanoate had been formed.

(a) What is the number of moles of water present at equilibrium? [1]

(b) No record appears to exist of the volume of the glass tubes which Berthelot and St Gilles used. Explain why this does not matter in this case. [1]

(c) Calculate the value for the equilibrium constant, K_c, for this reaction at 400 K. [3]

(d) (i) What is the significance of the size of this equilibrium constant? [2]

(ii) What would be implied by an equilibrium constant value of, say, 10^{-17}, as in the reaction:

$H_2(g) + Cl_2(g) \rightleftharpoons 2HCl(g)$ at 300 K? [1]

(iii) What would be implied by an equilibrium constant value of, say, 10^{59} dm^3 mol^{-1}, as in the reaction:

$$2NO(g) + 2CO(g) \rightleftharpoons N_2(g) + 2CO_2(g) \text{ at 570 K?}$$ [1]

22.6 Sulphur dichloride dioxide decomposes readily into sulphur dioxide and chlorine.

$$SO_2Cl_2(g) \rightleftharpoons SO_2(g) + Cl_2(g)$$

At 300 K and 1 atmosphere pressure and starting with an initial mass of 2.70 g of sulphur dichloride dioxide in a closed vessel of 500 cm^3 capacity, it was found that, at equilibrium 0.54 g of it had decomposed.
What is the value for the equilibrium constant, K_c, for this reaction at 300 K? [5]

22.7 An important reaction in the manufacture of sulphuric acid is;

$$2SO_2(g) + O_2(g) \rightleftharpoons 2SO_3(g); \Delta H = -197 \text{ kJ mol}^{-1}.$$

This reaction was extensively investigated by Bodenstein in 1905. In one of his experiments he found that, at 1000 K, equilibrium occurred when the partial pressures of the gases in the mixture were as follows:

$$p_{SO_2} = 0.456 \text{ atmospheres,}$$

$$p_{O_2} = 0.180 \text{ atmospheres,}$$

$$p_{SO_3} = 0.364 \text{ atmospheres.}$$

(a) What does the term 'partial pressure' mean? [2]
(b) Write an expression for the equilibrium constant, K_p, for this reaction in terms of partial pressures. [2]
(c) Use the partial pressures given above to calculate the value of K_p at 1000 K. [3]
(d) What would be the effects on the partial pressures of each of the gases, in each case, if the following changes were made to the system? Also state, in each case, how the value of K_p would be affected, if at all.
   (i) The partial pressure of oxygen is increased by 0.100 atmospheres, the total pressure remaining constant. [2]
   (ii) The total pressure of the system was reduced to 0.800 atmospheres. [2]
   (iii) The temperature of the system was lowered to 800 K. [2]
   (iv) Argon was added to the mixture but the total pressure was maintained at its original value. [2]

22.8 In 1960 it was reported that in a study of 100 g of sodium vapour at 10 atmospheres pressure and 1500 K, equilibrium was reached when 71.3 g of sodium atoms were present, the remainder consisting of diatomic molecules.
(a) Write a chemical equation for the equilibrium which existed. [2]

(b) Write an expression which could be used to find the value of the equilibrium constant, K_p, for this system. **[2]**

(c) Use the data to calculate the value of K_p for this reaction at 1500 K. **[3]**

22.9 **(a)** In each of the following examples, state whether the equilibrium will shift towards the left of the equation, towards the right of the equation or remain unchanged when the alteration in conditions described takes place. Explain, in each case, how you reach your decision.

(i) $C_2H_4(g) + H_2O(g) \rightleftharpoons C_2H_5OH(g)$ – the total pressure is doubled. **[2]**

(ii) $2CO(g) + O_2(g) \rightleftharpoons 2CO_2(g)$ – the total pressure is increased. **[2]**

(iii) $H_2(g) + I_2(g) \rightleftharpoons 2HI(g)$ – the total pressure is reduced to one quarter of its original value. **[2]**

(iv) $N_2O_4(g) \rightleftharpoons 2NO_2(g)$ – the total pressure is halved. **[2]**

(v) $3Fe(s) + 4H_2O(g) \rightleftharpoons Fe_3O_4(s) + 4H_2(g)$ – the total pressure is doubled. **[2]**

State and explain whether K_p would increase, decrease or stay the same as a result of the changes in conditions. **[2]**

(b) In each of the following examples, state whether K_c would increase, decrease or stay the same on making the change described. Explain, in each case, how you reach your decision.

(i) $N_2(g) + 3H_2(g) \rightleftharpoons 2NH_3(g); \Delta H = -92.2 \text{ kJ mol}^{-1}$ – the temperature is raised from 650 K to 720 K. **[3]**

(ii) $Ag_2CO_3(s) \rightleftharpoons Ag_2O(s) + CO_2(g); \Delta H = +81.6 \text{ kJ mol}^{-1}$ – the temperature is reduced from 450 K to 400 K. **[3]**

(iii) $CaCO_3(s) \rightleftharpoons CaO(s) + CO_2(g); \Delta H = +177.8 \text{ kJ mol}^{-1}$ – the temperature is reduced from 1200 K to 1000 K. **[3]**

(iv) $H_2(g) + I_2(g) \rightleftharpoons 2HI(g); \Delta H = +2.9 \text{ kJ mol}^{-1}$ – the temperature is increased from 500 K to 700 K. **[3]**

(v) $N_2(g) + O_2(g) \rightleftharpoons 2NO(g); \Delta H = -30.4 \text{ kJ mol}^{-1}$ – the temperature is decreased from 1500 K to 1100 K. **[2]**

22.10 For the following four examples use the data to decide whether the **forward reaction** is exothermic or endothermic. Explain your reasoning in each case.

(i) $N_2O_4(g) \rightleftharpoons 2NO_2(g)$

temperature/K	K_p/atm
350	3.89
400	47.9
450	347
500	1700
550	6030

[2]

(ii) $2SO_2(g) + O_2(g) \rightleftharpoons 2SO_3(g)$

temperature/K	K_p/atm^{-1}	
300	4.0×10^{24}	
500	2.5×10^{10}	
700	3.0×10^4	
1100	1.3×10^{-1}	[2]

(iii) $H_2(g) + CO_2(g) \rightleftharpoons H_2O(g) + CO(g)$

temperature/K	$\lg K_p$	
500	-2.11	
700	-0.91	
900	-0.22	
1100	$+0.16$	
1300	$+0.45$	[2]

(iv) $CH_3CO_2H(aq) \rightleftharpoons CH_3CO_2^-(aq) + H^+(aq)$

temperature/K	$K_a/10^{-5}\,\text{mol dm}^{-3}$	
293	1.75	
313	1.70	
333	1.63	[2]

22.11 Explain what is meant by:
(a) a strong acid [2]
(b) a weak acid [2]
(c) a strong alkali [2]
(d) a weak alkali. [2]

22.12 **(a)** Calculate the pH of the following solutions of strong acids assuming that they are completely dissociated into ions.
(i) hydrochloric acid, $HCl(aq)$, of concentration $0.01\,\text{mol dm}^{-3}$ [2]
(ii) sulphuric acid, $H_2SO_4(aq)$, of concentration $0.01\,\text{mol dm}^{-3}$ [2]
(iii) hydrobromic acid, $HBr(aq)$, of concentration $0.001\,\text{mol dm}^{-3}$ [2]
(iv) sulphuric acid, $H_2SO_4(aq)$, of concentration $0.005\,\text{mol dm}^{-3}$ [2]
(v) nitric acid, $HNO_3(aq)$, of concentration $0.005\,\text{mol dm}^{-3}$ [2]
(b) Calculate the pH of the following solutions of strong alkalis assuming that they are completely dissociated into ions.
(i) sodium hydroxide, $NaOH(aq)$, of concentration $0.01\,\text{mol dm}^{-3}$ [2]
(ii) barium hydroxide, $Ba(OH)_2(aq)$, of concentration $0.01\,\text{mol dm}^{-3}$ [2]
(iii) potassium hydroxide, $KOH(aq)$, of concentration $0.001\,\text{mol dm}^{-3}$ [2]

       (iv) sodium hydroxide, $NaOH(aq)$, of concentration $0.005 \text{ mol dm}^{-3}$ **[2]**

       (v) calcium hydroxide, $Ca(OH)_2(aq)$, of concentration
       $3 \times 10^{-4} \text{ mol dm}^{-3}(aq)$ **[2]**

22.13 Calculate the pH of the following solutions:

(a) sulphuric acid, $H_2SO_4(aq)$, of concentration 2.0 mol dm^{-3}, assuming that 50% of the acid is dissociated into ions   **[3]**

(b) hydrochloric acid, $HCl(aq)$, of concentration 5.0 mol dm^{-3}, assuming that 40% of the acid is dissociated into ions   **[3]**

(c) barium hydroxide, $Ba(OH)_2(aq)$, of concentration 3.0 mol dm^{-3}, assuming that 60% of the alkali is dissociated into ions   **[3]**

(d) potassium hydroxide, $KOH(aq)$, of concentration 10.0 mol dm^{-3}, assuming that it is 27% dissociated into ions.   **[3]**

22.14 **(a)** For the following solutions, convert their pH values into corresponding hydrogen ion concentrations:

       (i) hydrochloric acid, $HCl(aq)$, of pH 1.0   **[1]**

       (ii) sulphuric acid, H_2SO_4, of pH 3.0   **[1]**

       (iii) nitric acid, $HNO_3(aq)$, of pH 1.87   **[1]**

       (iv) sodium hydroxide, $NaOH(aq)$, of pH 12.0   **[1]**

       (v) barium hydroxide, $Ba(OH)_2(aq)$, of pH 9.34.   **[1]**

(b) What are the hydroxide concentrations in solutions **(a)(iv)** and **(a)(v)**?   **[1]**

22.15 Find the pH of the following solutions of weak acids using the K_a values shown:

(a) 2-hydroxypropanoic acid, $CH_3CHOHCO_2H(aq)$, of concentration $0.001 \text{ mol dm}^{-3}$, $(K_a = 1.4 \times 10^{-4} \text{ mol dm}^{-3})$ **[4]**

(b) nitrous acid, $HNO_3(aq)$ of concentration 0.30 mol dm^{-3}, $(K_a = 4.7 \times 10^{-4} \text{ mol dm}^{-3})$ **[3]**

(c) methanoic acid, $HCO_2H(aq)$, of concentration 0.01 mol dm^{-3}, $(K_a = 1.6 \times 10^{-4} \text{ mol dm}^{-3})$ **[3]**

(d) propanoic acid, $C_2H_5CO_2H(aq)$, of concentration 0.01 mol dm^{-3}, $(K_a = 1.3 \times 10^{-6} \text{ mol dm}^{-3})$.   **[3]**

22.16 Find the value of the acid dissociation constant, K_a, for each of the acids listed below, using the information given:

(a) the pH of a solution of ethanoic acid, $CH_3CO_2H(aq)$, of concentration 0.01 mol dm^{-3} is 3.37   **[4]**

(b) the pH of a solution of benzoic acid, $C_6H_5CO_2H(aq)$, of concentration 0.02 mol dm^{-3} is 2.94   **[3]**

(c) the pH of a solution of boric acid, $H_3BO_3(aq)$, of concentration 0.01 mol dm^{-3} is 5.60   **[3]**

(d) the pH of a solution of butanoic acid, $C_3H_7CO_2H(aq)$, of concentration 0.01 mol dm^{-3} is 3.42.   **[3]**

22.17 An experiment is carried out to measure K_a for ethanoic acid, CH_3CO_2H. 100 cm³ of sodium ethanoate solution of concentration 1.0 mol dm⁻³ is placed in a beaker and 20 cm³ of hydrochloric acid of concentration 1.0 mol dm⁻³ is added followed in turn by the addition of one drop of a universal indicator solution. The colour obtained is matched with the colour produced in a series of solutions of known pH. Three further 20 cm³ additions of the same acid are made and, after each addition, the pH is determined by the same technique.

The results obtained are:

after first addition of acid	pH 5.0
after second addition of acid	pH 4.9
after third addition of acid	pH 4.6
after fourth addition of acid	pH 4.3

(a) Write an equation for the reaction between sodium ethanoate and hydrochloric acid. **[1]**

(b) What name is given to the kind of solution formed in this experiment? **[1]**

(c) Calculate for each of the four additions a value for the acid dissociation constant, K_a, for ethanoic acid. [**4 marks for the first calculation, 1 mark for each subsequent calculation.**]

(d) Discuss the limitations of this method for determining K_a for a weak acid. **[2]**

(e) What value would you finally quote for K_a as a result of this experiment? **[1]**

22.18 **(a)** Calculate the pH of a buffer solution which is a mixture of equal volumes of ethanoic acid of concentration 0.10 mol dm⁻³ and of sodium ethanoate of concentration 0.40 mol dm⁻³. ($K_a = 1.7 \times 10^{-5}$ mol dm⁻³) **[2]**

(b) What is the pH of the buffer made by adding equimolar amounts of aqueous ammonia solution and aqueous ammonium chloride solution? (K_a for the ammonium ion, $NH_4^+(aq)$, $= 5.6 \times 10^{-10}$ mol dm⁻³) **[2]**

(c) 20.0 cm³ of a solution of methanoic acid, HCO_2H, of concentration 0.20 mol dm⁻³ is mixed with 40.0 cm³ of a solution of sodium methanoate of concentration 0.05 mol dm⁻³. What is the pH of the resulting mixture? (K_a of methanoic acid $= 1.6 \times 10^{-4}$ mol dm⁻³) **[2]**

(d) A solution is made by mixing 25.0 cm³ of aqueous potassium dihydrogenphosphate(V), KH_2PO_4, with 75.0 cm³ of aqueous dipotassium hydrogenphosphate(V), K_2HPO_4, both solutions being of concentration 0.05 mol dm⁻³. What is the pH of the mixture? (K_a for the dihydrogenphosphate(V) ion is 6.2×10^{-8} mol dm⁻³) **[2]**

(e) pK_a for the methylammonium ion, $CH_3NH_3^+$, is 3.34. Calculate the pH of a solution containing 0.01 moles of methylamine and 0.03 moles of methylamine hydrochloride in 500 cm³ of solution. **[2]**

(f) (i) The weak acid, hydrofluoric acid, HF, has a pK_a value of 3.30 at 298 K. If 100 cm^3 of a solution of this acid of concentration 0.50 mol dm^{-3} was mixed with 50 cm^3 of a solution of sodium fluoride of concentration 0.10 mol dm^{-3} what pH would the resulting mixture possess? **[2]**

(ii) Explain, in terms of the equilibria involved, what would happen if **(A)** a small quantity of a dilute solution of a strong acid was added to the mixture, and **(B)** a similar small quantity of a dilute solution of a strong alkali was added. **[5]**

Periodic trends in reactivity

- the reactions of elements with oxygen, chlorine and water
- the ionic or covalent structures of oxides and chlorides
- how the position of an element in the periodic table influences the type of bonding in its oxide and chloride
- the fact that ionic and covalent chlorides behave differently with water
- the acidic, basic or amphoteric character of oxides
- how the position of an element in the periodic table influences the acid–base character of its oxide.

23.1 **(a)** Arrange the elements sodium, magnesium and aluminium in order of increasing reactivity with water (least reactive first). [1]

(b) Suggest an explanation for the reactivity trend noted in **(a)**. [3]

23.2 **(a)** Under what conditions will magnesium react with water? [2]

(b) Write an equation for the reaction which occurs. [2]

23.3 **(a)** Give the formulae of the chlorides formed by aluminium, silicon and phosphorus. [4]

(b) Explain why phosphorus is the only one of these elements to form two chlorides. [3]

23.4 Sodium reacts with oxygen to give a product, X, containing 25.8% oxygen, which reacts further on heating with oxygen under pressure to give a compound Y of empirical formula NaO with a relative formula mass of 78.

(a) Calculate the formula of X. [2]

(b) **(i)** Calculate the formula of Y. [2]

(ii) Give the formula of the anion containing oxygen contained in Y and name it. [2]

(c) Suggest why lithium oxide, Li_2O, fails to react with oxygen to give the lithium equivalent of compound Y. [3]

23.5* **(a)** Write an equation for the complete hydrolysis of phosphorus(V) chloride. [2]

(b) The hydrolysis product containing phosphorus is a tribasic acid. What is implied by the term *tribasic* and how can this behaviour be explained in terms of its structure? [3]

23.6 Phosphorus(III) chloride is a liquid which does not conduct electricity, but on addition to water it gives a solution which is a good electrical conductor. Explain these observations. [9]

23.7* Phosphorus forms a trichloride, PCl_3 and a pentachloride, PCl_5, but sulphur forms no pentachloride, and its tetrachloride decomposes above $-31\ °C$. Suggest explanations for the differing behaviour of these two elements. [4]

23.8 **(a) (i)** Give the formulae of the two oxides of sulphur. [2]
   (ii) State briefly how each oxide can be prepared. [4]
  (b) Write balanced equations for the reaction of each oxide with water, naming the products formed. [4]

23.9 **(a) (i)** Write a balanced equation for the reaction of chlorine with water at room temperature. [2]
   (ii) Write an equation for the reaction which occurs when the reaction mixture from **(i)** is boiled. [2]
  (b) Name the oxoanions formed in each reaction and give their shapes. [4]

23.10 Fluorine reacts with water to liberate oxygen gas.
  (a) Write an equation for the reaction between water and fluorine. [2]
  (b) Using the concept of oxidation states, show that a redox reaction has taken place. [4]

23.11 For each of the following elements give the formula of its oxide and state the main type of bonding present in the oxide: sodium, magnesium, aluminium, phosphorus, sulphur. [5]

23.12 **(a)** What name is given to oxides which react with both acids and alkalis? [1]
  (b) For each of the following elements, give the formula and acid–base character of one of its oxides: potassium, barium, aluminium, nitrogen and chlorine. [5]

23.13 Some oxides dissolve in acids, some dissolve in alkalis and a few dissolve in both. Write equations to summarize the behaviour of the following with acids and alkalis: MgO, Al_2O_3, SiO_2. [10]

23.14 Write the formulae of all the chlorides of the following elements, stating in each case whether the chloride reacts with water or not:
sodium, magnesium, aluminium, phosphorus and sulphur. [8]

23.15 What trend is evident in the type of bonding in both oxides and chlorides across period 3? Explain this trend. [4]

23.16* Explain the difference in the melting points of carbon dioxide ($-57\ °C$) and silicon dioxide ($1700\ °C$) in terms of their bonding. [6]

23.17 In which of the following is the bonding largely covalent:
CO_2, MgO, $AlCl_3$, KCl, SiO_2, HCl? [4]

23.18 Aluminium chloride sublimes at around 200 °C. What does this suggest about the type of bonding present in this chloride? Suggest an explanation in terms of the position of aluminium in the periodic table.   [3]

23.19 State the type of bonding present in each of the following chlorides. Write equations for those which react with water: NaCl, CCl$_4$, PCl$_5$, AlCl$_3$, HCl.   [11]

23.20 Solid aluminium chloride has a relative molecular mass of 267 but the relative molecular mass of the vapour at around 200 °C is half this value. Suggest a structural explanation for these facts.   [4]

24 The Transition Elements

In this section you will need to understand:

- how to predict the electronic structure of a transition element from a knowledge of its atomic number
- how electrons are lost from transition metal atoms to form ions
- the fact that transition metals form at least one ion having an incomplete d-orbital
- the existence of more than one stable oxidation state for transition metals
- the colour of many transition metal compounds and ions in solution
- the fact that transition metal ions can accept lone pairs of electrons from ligands to form complexes
- the catalytic properties of many transition metals and their compounds.

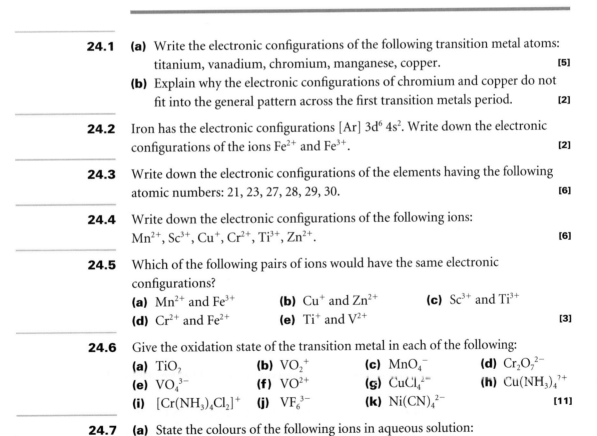

24.1 **(a)** Write the electronic configurations of the following transition metal atoms: titanium, vanadium, chromium, manganese, copper. **[5]**

(b) Explain why the electronic configurations of chromium and copper do not fit into the general pattern across the first transition metals period. **[2]**

24.2 Iron has the electronic configurations [Ar] $3d^6$ $4s^2$. Write down the electronic configurations of the ions Fe^{2+} and Fe^{3+}. **[2]**

24.3 Write down the electronic configurations of the elements having the following atomic numbers: 21, 23, 27, 28, 29, 30. **[6]**

24.4 Write down the electronic configurations of the following ions: Mn^{2+}, Sc^{3+}, Cu^+, Cr^{2+}, Ti^{3+}, Zn^{2+}. **[6]**

24.5 Which of the following pairs of ions would have the same electronic configurations?

(a) Mn^{2+} and Fe^{3+} **(b)** Cu^+ and Zn^{2+} **(c)** Sc^{3+} and Ti^{3+}

(d) Cr^{2+} and Fe^{2+} **(e)** Ti^+ and V^{2+} **[3]**

24.6 Give the oxidation state of the transition metal in each of the following:

(a) TiO_2 **(b)** VO_2^+ **(c)** MnO_4^- **(d)** $Cr_2O_7^{2-}$

(e) VO_4^{3-} **(f)** VO^{2+} **(g)** $CuCl_4^{2-}$ **(h)** $Cu(NH_3)_4^{2+}$

(i) $[Cr(NH_3)_4Cl_2]^+$ **(j)** VF_6^{3-} **(k)** $Ni(CN)_4^{2-}$ **[11]**

24.7 **(a)** State the colours of the following ions in aqueous solution:

(i) Fe^{3+} **(ii)** Fe^{2+} **(iii)** Cu^{2+}

(iv) Zn^{2+} **(v)** MnO_4^- **(vi)** $Cr_2O_7^{2-}$ **[6]**

(b) Explain briefly why solutions of many transition metal ions appear coloured in aqueous solution. **[3]**

24.8 What is meant by the terms 'complex ion' and 'ligand'? **[4]**

24.9 Write the formulae of **three** ligands in each case which:
(a) are neutral **[3]**
(b) have a negative charge. **[3]**

24.10 **(a)** Write down the formula of the complex ions which match the following requirements:

metal and its oxidation state	ligands
Cu(I)	four Cl
Ag(I)	two NH_3
Ti(IV)	six Cl
Co(III)	four NH_3 + two H_2O
Cr(III)	four NH_3 + two Cl

[5]

(b) Sketch and give the shape of each complex ion. **[11]**

24.11 Many transition metals ions in aqueous solution give precipitates on the addition of aqueous sodium hydroxide solution.
(a) For each of the following ions, give the colour of the precipitate formed with aqueous sodium hydroxide solution: Fe^{2+}, Fe^{3+}, Mn^{2+}, Cr^{3+}, Ni^{2+}, Co^{2+}. **[6]**
(b) The formation of such precipitates involves the deprotonation of the aquo cations. Write an equation for the reaction between $Fe^{2+}(aq)$ and the $OH^-(aq)$ to illustrate the meaning of the term 'deprotonation'. **[2]**
(c) Only one of the precipitates in **(a)** dissolves in excess sodium hydroxide solution. Write an equation for the reaction which occurs when the precipitate dissolves. **[3]**
(d) Which of the precipitates dissolves in excess aqueous ammonia solution? **[3]**

24.12 The addition of concentrated hydrochloric acid to blue aqueous solutions of copper(II) ions leads to the formation of a green solution containing a yellow complex of copper(II) with chloride ions. The subsequent addition of concentrated ammonia solution leads first to the formation of a pale-blue precipitate and finally a deep-blue solution.
(a) **(i)** Suggest a formula for the yellow complex.
(ii) Write a balanced equation to represent the formation of the yellow complex.
(iii) What type of reaction does this equation represent? **[4]**
(b) **(i)** Write down the formula of the pale-blue precipitate.
(ii) Write a balanced equation to show the formation of the deep-blue solution from this precipate. **[4]**

24.13 Transition metals and their compounds are important industrial catalysts.

(a) Identify an industrial process which uses the following metal, or one of its compounds as a catalyst: vanadium, iron and nickel. [3]

(b) Write down an equation for the reaction catalysed industrially using (i) a vanadium compound (ii) iron. [4]

(c) What property of the transition metals is likely to be most important in enabling them, or their compounds, to act as catalysts? [1]

Shapes of organic molecules

25

- the shapes of organic compounds
- the different bond angles in organic compounds
- optical isomerism.

25.1 **(a)** What is the approximate H—C—H bond angle in:

   (i) methane

   (ii) propane

   (iii) ethene? [4]

(b) What is the C—C—C bond angle in:

   (i) propane

   (ii) propene

   (iii) propyne

   (iv) benzene

   (v) cyclohexane? [5]

25.2 What are the approximate bond angles marked in each of the following molecules? [9]

(a)
$$CH_3 \underset{\text{(i)}}{\overset{\text{(ii)}}{-}} CH_2 - O - H$$

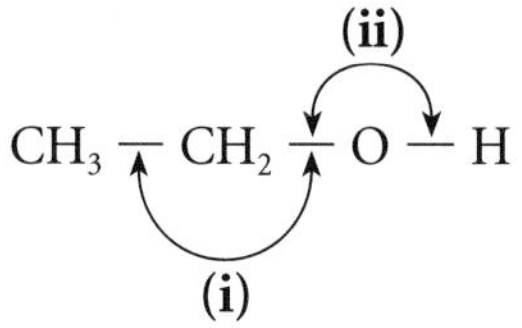

(b)
$$CH_3 \overset{\text{(i)}}{-} C \overset{}{-} CH_3$$
(ii) O

(c)
$$CH_3 \overset{\text{(i)}}{-} C \overset{}{-} O - H$$
(ii) O (iii)

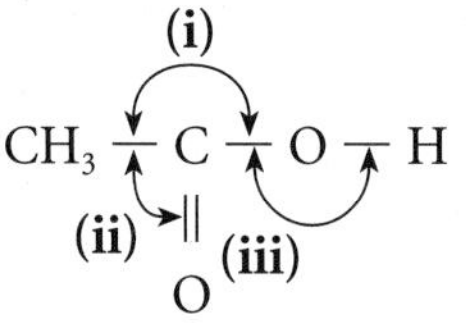

(d)
$$CH_3 \overset{\text{(i)}}{-} C \overset{}{-} H$$
(ii) O

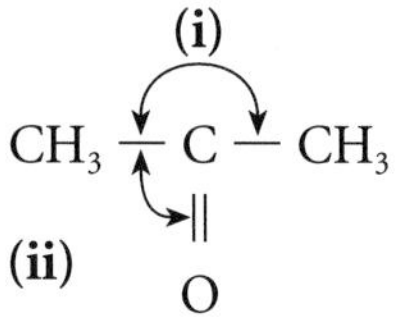

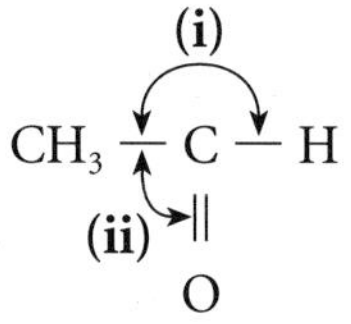

25.3 **(a)** Place the four bonds angles P, Q, R, S in order starting with the smallest. [1]

(b) The bonds in four carbonyl containing compounds are shown. Each is close to 120°, but will vary slightly, depending upon the size of the groups attached to the carbon atom.

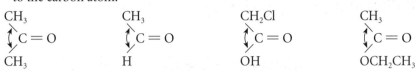

Starting with the smallest marked bond angle write the names of the four compounds. [5]

25.4* Give formulae and names of each compound.

(a) A tetrahedral molecule which contains:

    (i) 1 C atom

    (ii) 5 C atoms. [2]

(b) A 6 C alkene which is planar. [1]

(c) A planar molecule which contains only 2 carbon atoms. [1]

(d) A linear molecule which contains only 2 carbon atoms. [1]

(e) Cyclohexane is a molecule which is sometimes a 'boat' shape and at other times a 'chair' shape.

    (i) Draw the two structures. [2]

    (ii) State which shape is sterically more likely and explain why. [1]

(f) Buta-1,3-diene can exist in two stereo forms, *cis* and *trans*.

    (i) Draw the two forms. [2]

    (ii) Explain why in 1,1,4,4-tetrachlorobuta-1,3-diene the *trans* form is more stable than the *cis* form. [1]

25.5 The descriptions of some shapes of molecules, or parts of molecules, are given below:

A   regular trigonal

B   strongly distorted trigonal

C   linear

D   regular tetrahedral

E   slightly distorted tetrahedral

F   strongly distorted tetrahedral.

For the bonds around the carbon in each 'functional group or situation' select the appropriate name for their spatial distribution from the list above.

For example, the distribution in the carbonyl group in propanone is type A because the carbonyl group, C=O has two methyl groups bonded to it, and the C—C—C bond angle is just less than 120°.

(a) The carbonyl group in ethanoic acid.

(b) The carbonyl group in sodium ethanoate.

(c) The alkene bond in but-2-ene.

(d) The carbon, carbon, oxygen bond angle in propan-2-ol.

(e) The carbon, carbon, chlorine bond angle in chlorobenzene.

(f) The carbon, carbon, carbon bond angle in 2,2-dimethylpropane.

(g) The carbon, carbon, carbon bond angle in 2,2-dichloropropane. [7]

26 Arenes

In this section you will need to understand:

- how to name aromatic ring compounds
- the electrophilic substitution mechanism of the benzene ring
- the factors which affect substitution of the benzene ring
- the oxidation reactions of substituents on the ring
- some of the reactions of aromatic compounds.

26.1 **(a)** Write the names of the following arenes:

(i)  C₆H₅—CH₃

(ii)  C₆H₅—CH₂CH₃

(iii) 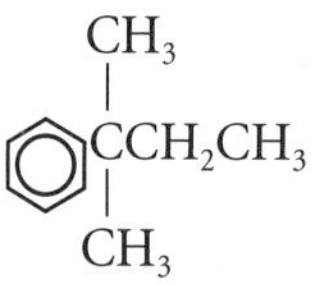 $C_6H_5-C(CH_3)_2CH_2CH_3$ (ring with CCH₂CH₃, CH₃ above and CH₃ below)

(iv)  C₆H₅—NH₂

(v)  C₆H₅—COOH

(vi)  C₆H₅—CHO

(vii)  C₆H₅—CCH₃ with ‖O (acetophenone)

(viii) C₆H₅—OH   [8]

(b) Which of the above arenes will form benzoic acid when reacted with alkaline potassium manganate(VII)?   [5]

26.2 Write the formulae for:
(a) 2-methylphenol
(b) 2,4-dimethylbenzoic acid
(c) methylbenzene (toluene)
(d) 2,4,6-trinitromethylbenzene (TNT)
(e) benzamide
(f) phenylamine (aminobenzene).   [6]

26.3 Complete each equation by giving the formula and name of the simplest organic product formed in each reaction:

(a)  C₆H₆ + HNO_3/H_2SO_4 →

(b)  C₆H₅NO₂ + HCl/Sn →

(c)  C₆H₆ + $CH_3Cl/AlCl_3$ →

(d)  C₆H₅CH₂CH₃ + $KMnO_4/OH^-$ →

(e)  C₆H₅NH₂ + $NaNO_2/HCl$ →   [10]

26.4* **(a)** What is the name and formula of the product in each of the following:

(i)  $\bigcirc$ + $Br_2/AlBr_3$ →

(ii)  $\bigcirc$ + $Cl_2/AlCl_3$ →

(iii)  $\bigcirc$ + $CH_3CH_2Cl/AlCl_3$ →

(iv)  $\bigcirc$ + $CH_3COCl/AlCl_3$ →

(v)  $\bigcirc$ + $CH_3CH_2COCl/AlCl_3$ → **[10]**

(b) Each of the following Friedel–Crafts reactions occurs in high yields overall, but in specific positions on the benzene ring.

(i) Explain why there is a 98% yield of a mixture of two isomers for:

$$CH_3CH_2\bigcirc + CH_3COCl \xrightarrow{AlCl_3}$$ **[4]**

(ii) Explain why there is a 96% yield of one isomer for:

$$NO_2\bigcirc + Br_2 \xrightarrow{AlBr_3}$$ **[4]**

26.5 **(a)** Give the mechanism for the substitution of a nitro group onto a benzene ring. **[5]**

(b) What reagent is used to convert nitrobenzene into aminobenzene? **[1]**

(c) Describe the reaction conditions needed to form a diazonium salt from aminobenzene. **[1]**

(d) Write the formulae for products obtained when benzenediazonium chloride is added to:

(i) phenol **[1]**

(ii) 2-naphthol. **[1]**

26.6 **(a)** Explain why side chains or functional groups which donate electrons to the benzene ring make it easier for electrophilic substitution to occur. **[5]**

(b) Which of the following groups increase the reactivity of a benzene ring: **[3]** CH_3, OH, NO_2, NH_2, $COOH$?

27 Reactions of aldehydes and ketones

In this section you will need to understand:

- the reduction of the carbonyl group
- addition reactions of the carbonyl group
- condensation reactions of aldehydes and ketones
- oxidation reactions of aldehydes
- chemical methods of distinguishing carbonyl groups and the difference between aldehydes and ketones.

27.1 **(a)** Write the formula for the organic product of the reaction (if any) of butanone with:

    (i) HCN

    (ii) $(NO_2)_2$⬡$NHNH_2$ (2,4-dinitrophenylhydrazine)

    (iii) $[Ag(NH_3)_2]^+$

    (iv) $NaBH_4$

    (v) $LiAlH_4$ **[5]**

(b) For the same series of reagents shown in **(a)(i)–(v)**, write the formula of the product, if formed, when butanal is reacted with each. **[5]**

27.2 A compound gives a positive result when treated with 2,4-dinitrophenylhydrazine, and a positive result with Tollens' reagent.

(a) What functional group is present? **[1]**

(b) Which other reagent could have been used to confirm the presence of this group? **[1]**

(c) By what physical method can the presence of this functional group be detected? **[1]**

27.3 Describe any changes when each of the following is warmed with either Fehling's reagent or ammoniacal silver ions, Tollens' reagent. Also name and give the formula of any organic product which is formed.

(a) CH_3CH_2CHO

(b) CH_3CCH_3
       ||
       O

(c) CH_3CHCH_3
       |
       OH

(d) $CH_3CH_2CH_2OH$

(e)  ⬡CHO

(f) CH_3CHO **[9]**

27.4   Aldehydes can be reduced to alcohols with a number of different reagents.

(a) What is the correct order for the following three statements to ensure the reduction occurs?   **[1]**

A   Add, in small portions, lithium tetrahydridoaluminate(III) powder.

B   Dissolve the aldehyde in dry ethoxyethane.

C   Fractionally distil ethoxyethane and ethanol.

(b) Write an equation for the reaction which will occur if the method described in **(a)** is carried out.   **[1]**

(c) Name another reagent which could have been used to reduce the aldehyde, and write its formula.   **[2]**

(d) Write the formula for lithium tetrahydridoaluminate(III).   **[1]**

27.5   What are the names of the organic products formed in the following reactions?

(a) HCHO $+ NaBH_4 \rightarrow$

(b) HCHO $+H^+/Cr_2O_7^{2-} \rightarrow$

(c) $CH_3COCH_3 + NaBH_4 \rightarrow$

(d)  ⬡$COCH_3 + NaBH_4 \rightarrow$

(e)  ⬡CHO $+ NaBH_4 \rightarrow$

(f)  ⬡CHO $+ H^+/Cr_2O_7^{2-} \rightarrow$ **[6]**

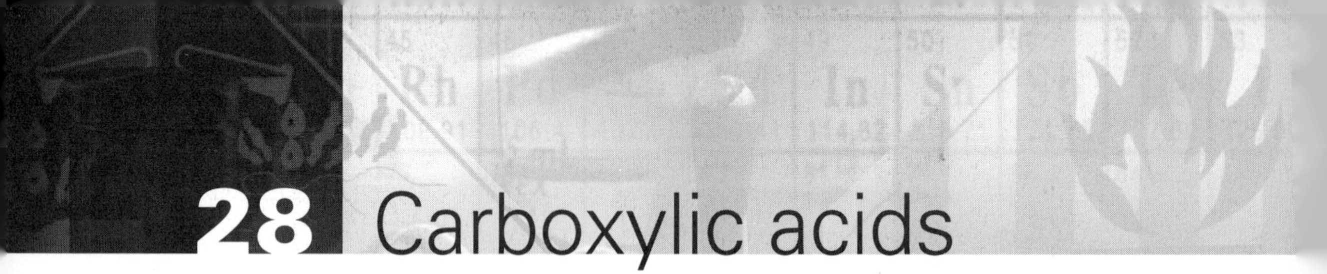

28 Carboxylic acids

In this section you will need to understand:

- the formation of acid chlorides
- the formation of esters
- the formation of amides
- the reduction of carboxylic acids to alcohols
- the mechanism for esterification.

28.1 **(a)** What are the names of:

    (i) CH_3CH_2COOH **(ii)** $HCOOH$

    (iii) COOH   **(iv)**  ⬡COOH
         |
         COOH

    (v)  ⬡CH_2COOH **(vi)**  HO⬡$COOH$ **[6]**

  (b) Write the formulae for the methyl esters formed from each of the acids in **(a)(i)–(vi)**.   **[7]**

28.2 The bond length for $C=O$ in propanone is 0.122 nm and that for $C—O$ in ethanol is 0.143 nm. The bond length for the $C=O$ in ethanoic acid appears to be the same as that of the $C—O$ bond and intermediate in length between the $C=O$ and $C—O$ values. Explain these data.   **[4]**

28.3 **(a)** Carboxylic acids are neutralized by concentrated ammonia solution. The salt, when heated, forms an amide. Write the two-step equation for the formation of propanamide.   **[3]**

  (b) An alternative method is to form the acid chloride and react this with concentrated ammonia solution. Write this two-step equation.   **[3]**

  (c) Suggest why the acid chloride route gives a better yield of amide.   **[1]**

28.4 Write the formula for each of the following and state what the products of hydrolysis are:

  (a) **(i)** ethyl ethanoate   **[3]**
      (ii) methyl ethanoate   **[3]**
      (iii) ethyl methanoate   **[3]**

  (b) ethyl benzoate   **[3]**

  (c) propanoyl chloride.   **[3]**

28.5 Esterification of ethanol and ethanoic acid is an equilibrium reaction. In the presence of concentrated sulphuric acid a yield of 80% can be obtained. Use curly arrows to show the mechanism for the reaction and describe the roles of the concentrated H_2SO_4. **[5]**

28.6* Ester formation by the reaction of alcohol with carboxylic acid is an equilibrium reaction. The presence of concentrated sulphuric acid favours the production of the ester. Ester formation by the reaction of an alcohol with an acid chloride in the presence of a base goes to completion. Explain, giving appropriate equations, why one reaction is an equilibrium and the other goes to completion. **[5]**

29 Reactions of amines and amides

In this section you will need to understand:

- substitution reactions to form amines
- dehydration of amides to form amines
- formation of amides by condensation reactions
- hydrolysis of nitriles to amides and carboxylic acids.

29.1 **(a)** Write the formula of:
    (i) 1-propylamine
    (ii) 2-propylamine (2-aminopropane)
    (iii) 2-aminopropanoic acid
    (iv) cyclohexylamine. [4]
  (b) For **(a)(iii)** show both stereoisomers. [2]
  (c) Write the formulae for:
    (i) 2-aminopropanamide
    (ii) 1,6-diaminohexane
    (iii) hexane-1,6-diamide
    (iv) benzene-1,4-dicarboxamide. [4]

29.2* Give the formula and name of each product from the following reactions:

  (a) $CH_3CONH_2 \xrightarrow{P_2O_5}$ [2]

  (b)  ⬡$CONH_2 \xrightarrow{P_2O_5}$ [2]

  (c) $H_2NOCCH_2CH_2CH_2CH_2CONH_2 \xrightarrow{P_2O_5}$ [2]

  (d) $CH_3C\equiv N \xrightarrow{LiAlH_4}$ [2]

  (e)  ⬡$CH_2C\equiv N \xrightarrow{LiAlH_4}$ [2]

29.3* Nitriles can be hydrolysed by an acid or a base. The first stage occurs in mild conditions but the second step requires heating under reflux.
  (a) Write two-stage equations for the acid hydrolysis of:
    (i) ethanenitrile [2]
    (ii) benzonitrile. [2]
  (b) What is the difference in the products formed if alkaline hydrolysis is used instead of acidic? [1]

29.4* Benzenediazonium chloride is only stable in the cold. If it is warmed it reacts with water to form phenol.

    (a) Show the steps and reagents and conditions in forming phenol from aminobenzene.   **[2]**

    (b) Suggest what is formed when benzylamide is treated with $NaNO_2/HCl$ in the cold and allowed to warm.   **[1]**

    (c) (i) Write the formula of the product formed when phenol is added to cold benzenediazonium chloride.   **[1]**

        (ii) In the Kekulé structure of a benzene ring the three double bonds are seen to be conjugated. How many double bonds are conjugated in the product from **(c)(i)**?   **[1]**

29.5* **(a)** Describe **two** routes to make 1-aminopropane from propanoic acid.   **[4]**

    (b) Describe how to make 1-aminopropane from propan-1-ol.   **[2]**

    (c) In what form does 1-aminopropane exist in hydrochloric acid?   **[1]**

    (d) (i) Write the formula of 2-aminopropanoic acid.   **[1]**

        (ii) Show the amino acid as a zwitterion.   **[1]**

Synthetic pathways

- two stage routes to form the following different product groups:
    halogenoalkanes
    carboxylic acids
    alcohols
    esters
    aldehydes and ketones
    amides
    amines
- multi-step reaction pathways.

30.1 Two-step syntheses of halogenoalkanes
Give the **formula** of the missing reactant or compound for the following two-step syntheses:

(a)

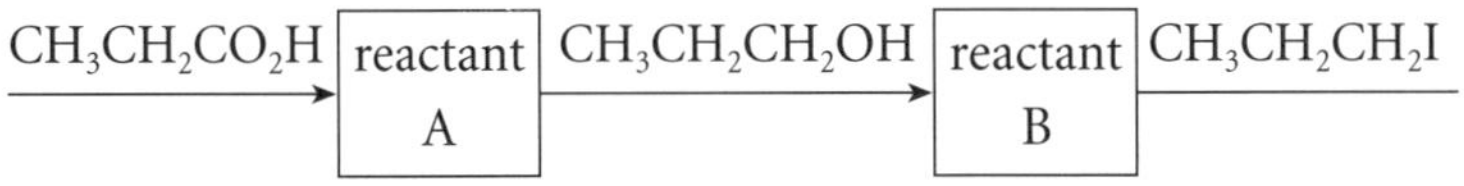

$$CH_3CH_2CO_2H \xrightarrow{\boxed{\begin{array}{c}\text{reactant}\\A\end{array}}} CH_3CH_2CH_2OH \xrightarrow{\boxed{\begin{array}{c}\text{reactant}\\B\end{array}}} CH_3CH_2CH_2I$$

[2]

(b)

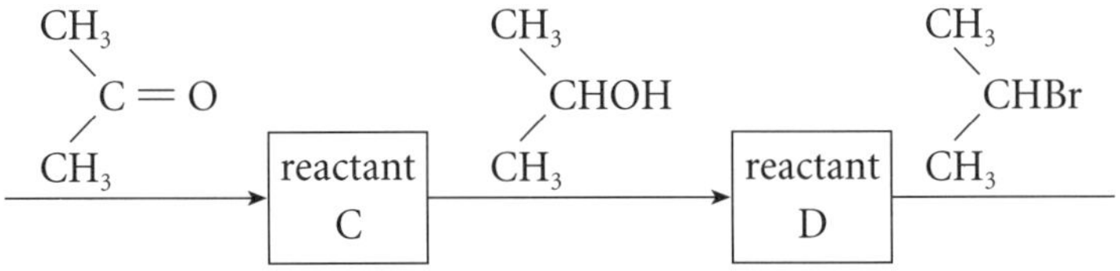

$$\begin{array}{c}CH_3\\ \diagdown\\ C=O\\ \diagup\\ CH_3\end{array} \xrightarrow{\boxed{\begin{array}{c}\text{reactant}\\C\end{array}}} \begin{array}{c}CH_3\\ \diagdown\\ CHOH\\ \diagup\\ CH_3\end{array} \xrightarrow{\boxed{\begin{array}{c}\text{reactant}\\D\end{array}}} \begin{array}{c}CH_3\\ \diagdown\\ CHBr\\ \diagup\\ CH_3\end{array}$$

[2]

(c)

$$\text{C}_6\text{H}_5\text{CH}_2\text{CH}_2\text{Cl} \xrightarrow{\boxed{\begin{array}{c}\text{reactant}\\E\end{array}}} \text{compound F} \xrightarrow{\boxed{\begin{array}{c}\text{reactant}\\G\end{array}}} \begin{array}{c}\text{C}_6\text{H}_5\text{CHCH}_3\\ |\\ \text{Cl}\end{array}$$

$$\text{C}_6\text{H}_5\text{CH}_2\text{CH}_2\text{OH} \xrightarrow{\boxed{P_2O_5}}$$

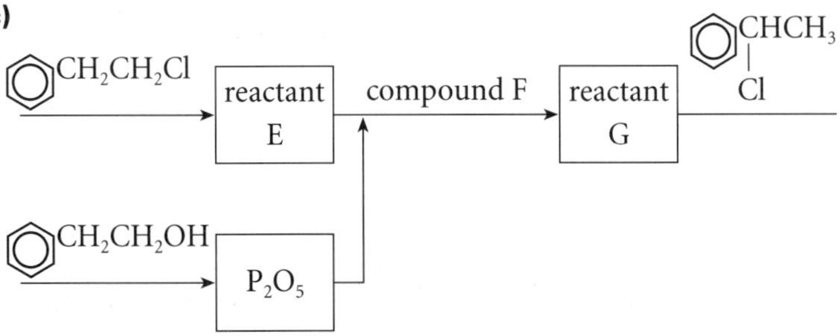

[3]

(d)

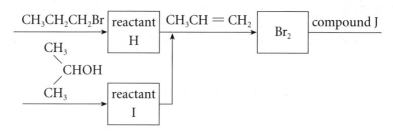

[3]

30.2* Two-step syntheses of carboxylic acids.
Give the **formula** of the missing reactant or compound for the following two-step syntheses:

(a)

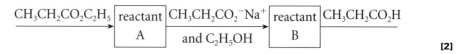

[2]

(b)

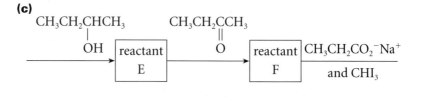

[2]

(c)

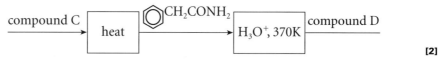

[2]

(d)

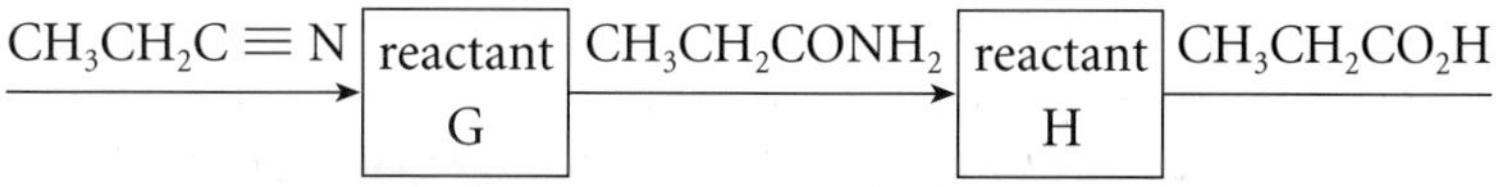

CH₃CH₂C≡N │reactant│ CH₃CH₂CONH₂ │reactant│ CH₃CH₂CO₂H
                    G   H

[2]

(e)

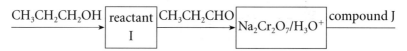

[2]

30.3 Two-step syntheses of alcohols.

Give the **formula** of the missing reactant or compound for the following two-step syntheses:

(a)

CH3CH2CHO
reactant A
CH3CH2COOH
reactant B
CH3CH2CH2OH

[2]

(b)

CH3CH = CH2
reactant C
CH3 CHBr CH3
reactant D
CH3 CHOH CH3

[2]

(c)

NH2
reactant E
N2+Cl−
reactant F
OH

[2]

(d)

CO2H
reactant G
CO2CH3
LiAlH4
compounds H and I

[3]

30.4 Two-step syntheses of esters.

Give the **formula** of the missing reactant or compound for the following two-step syntheses:

(a)

CH3CHC≡N
OH
reactant A
CH3CHCO2H
OH
reactant B
CH3CHCO2CH2CH3
OH

[2]

(b)

CH3CH2CH2OH
reactant C
compound D
catalyst E
CH3CH2CO2CH
CH3
CH3
CH3
CHOH
CH3
compound F
LiAlH4

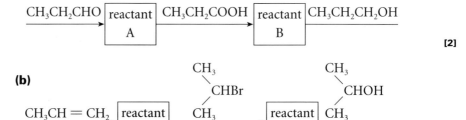

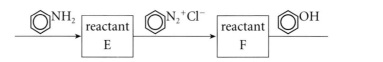

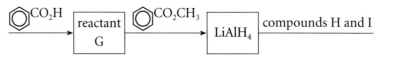

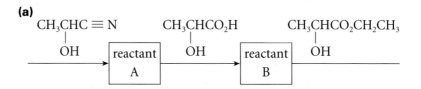

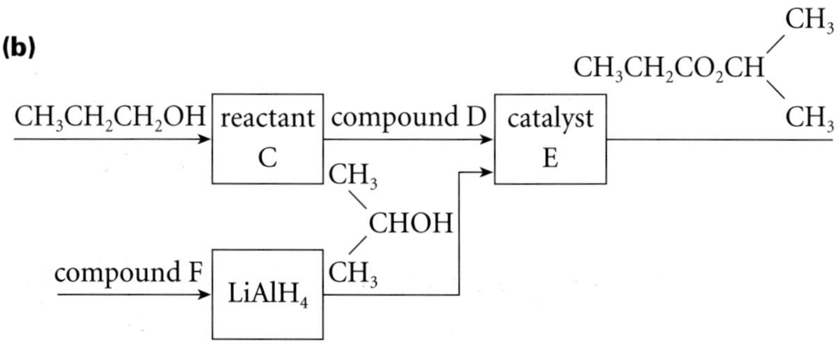

[4]

(c)

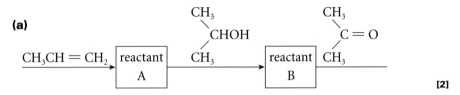

[2]

30.5 Two-step syntheses of aldehydes and ketones.
Give the **formula** of the missing reactant or compound for the following two-step syntheses:

(a)

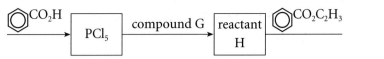

[2]

(b)

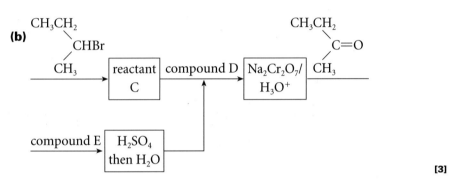

[3]

(c)

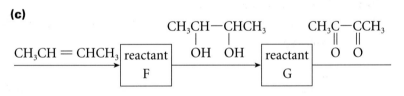

[2]

30.6* Two-step syntheses of amines.
Give the **formula** of the missing reactant or compound for the following two-step syntheses:

(a)

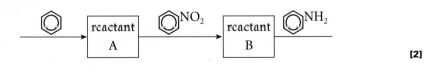

[2]

(b)

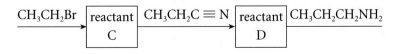

[2]

(c)

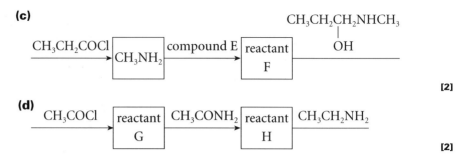

[2]

(d)

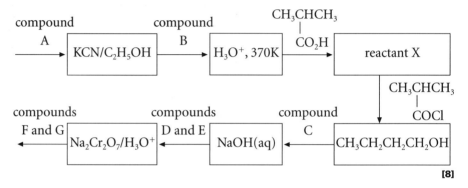

[2]

30.7 In your answer write the compound or reactant letter and the formula of the missing substance.

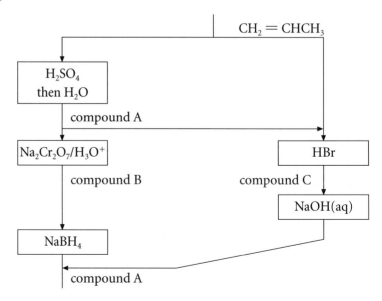

[8]

30.8 In your answer write the compound or reactant letter and the formula of the missing substance.

30.9* In your answer write the compound or reactant letter and the formula of the missing substance.

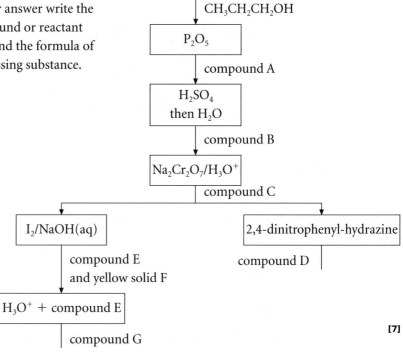

CH$_3$CH$_2$CH$_2$OH

P$_2$O$_5$

compound A

H$_2$SO$_4$ then H$_2$O

compound B

Na$_2$Cr$_2$O$_7$/H$_3$O$^+$

compound C

I$_2$/NaOH(aq)

compound E and yellow solid F

H$_3$O$^+$ + compound E

compound G

2,4-dinitrophenyl-hydrazine

compound D

[7]

30.10 In your answer write the compound or reactant letter and the formula of the missing substance.

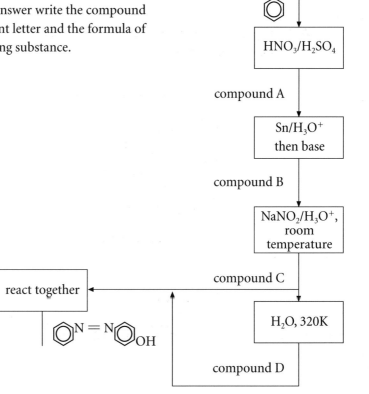

HNO$_3$/H$_2$SO$_4$

compound A

Sn/H$_3$O$^+$ then base

compound B

NaNO$_2$/H$_3$O$^+$, room temperature

compound C

react together

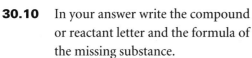

H$_2$O, 320K

compound D

[4]

30.11 In your answer write the compound or reactant letter and the formula of the missing substance.

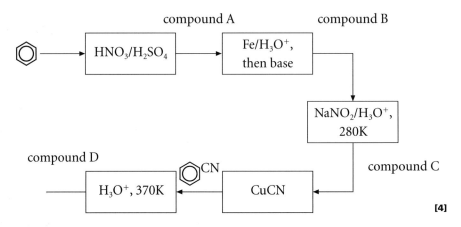

[4]

30.12 In your answer write the compound or reactant letter and the formula of the missing substance.

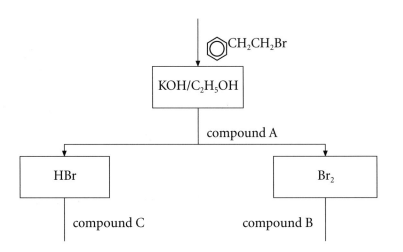

[3]

30.13* In your answer write the compound or reactant letter and the formula of the missing substance.

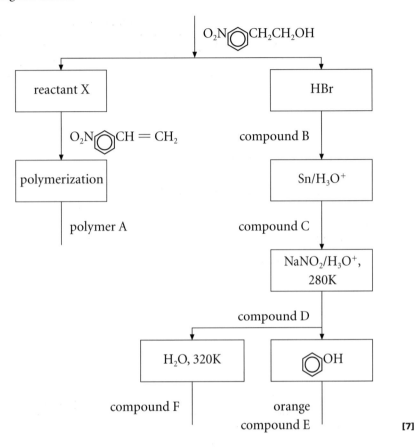

[7]

30.14 In your answer write the compound or reactant letter and the formula of the missing substance.

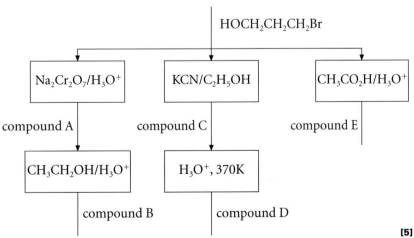

[5]

31 Polymerization

In this section you will need to understand:

- the mechanism of addition polymerization
- the structure and properties of various addition polymers
- the mechanism of condensation polymerization
- the structure and properties of various condensation polymers.

31.1* **(a)** Classify each of the following polymers into either addition, condensation polyester or condensation polyamide:

polypropene, terylene, nylon, acrylic, perspex, protein, polythene, rubber, PVA, PTFE, PVC polystyrene. [12]

(b) Draw the repeat units for:

polypropene, polystyrene, polythene, nylon 6,6, nylon 6, terylene. [6]

31.2 **(a)** Explain why:

(i) proteins and nylon 6 can be hydrolysed by acid [2]

(ii) terylene can be hydrolysed by alkali [2]

(iii) polythene is stable to hydrolysis. [1]

(b) What is formed by the acid hydrolysis of nylon 6? Give the name and formula of the product. [2]

31.3 Addition polymers, such as polypropene, can be classified as follows:

atactic – side chains are randomly orientated along chain; they are made by free radical polymerization.

isotactic – side chains along same side of chain; they are made using Ziegler–Natta catalysts.

syndiotactic – side chains alternate from side to side of the polymer chain; they require Ziegler–Natta catalysts and carefully controlled conditions.

(a) Which type of polypropene is probably the cheapest to product? Suggest why. [1]

(b) Suggest a use for each type of polypropene, and give an example of an object it could be used for. [6]

(c) Draw the syndiotactic form of polypropene. [2]

(d) Write a free radical mechanism for the propagation stages of the formation of isotactic polypropene. Show four monomers being involved. [3]

Testing for functional groups

- how to test for the following functional groups:
  - aldehydes
  - ketones
  - hydroxyl groups
  - halides
  - acids
  - alkenes
  - methyl ketones.

32.1 Substance A, is oxidized with acidified potassium dichromate(VI) and the product, B, is distilled off as it forms. The product, B, produces a silver mirror with Tollens' reagent.

(a) What type of compound is:
  (i) substance A   (ii) substance B?   [2]

(b) How does substance A react with sodium? What type of compound is formed?   [2]

(c) What type of compound is formed when substance A is oxidized under reflux?   [1]

32.2 Compound X (relative molecular mass 46) is neutral. It reacts with PCl_5. Compound Y (relative molecular mass 60) is acidic. It also reacts with PCl_5.

(a) What are the structural formulae of X and Y?   [2]

(b) Write a balanced equation for the reaction of compound X with compound Y.   [2]

(c) Explain why using the *product* of compound Y with PCl_5 will give a better yield when reacted with compound X than using compound Y.   [2]

32.3 Compound A was analysed. It turned bromine water colourless and after hydrolysis gave a cream precipitate with acidified silver nitrate solution. It had a relative molecular mass of 135.

(a) Write the empirical formulae of compound A.   [1]

(b) (i) Write the eight structural isomers.   [8]
   (ii) By each isomer, which may exist in a *cis* or *trans* form, write *cis/trans*.   [3]
   (iii) By the isomer, which may exist as an optical isomer, write *optical*.   [1]

32.4 A white solid burns with a very smoky flame. Its empirical formula is $C_7H_6O_2$.

    (a) What is its structural formula? [1]

    (b) What will it form with:

        (i) PCl_5 **(ii)** $NaHCO_3$ solution? [2]

    (c) Why does it burn with a smoky flame? [1]

32.5 A compound contains C, H and O in the ratio 1:3:1. On reacting with sodium, two of the H's are replaced by Na. The compound does not react with 2,4-dinitrophenylhydrazine.

    (a) What is the structural formula of the compound? [1]

    (b) What is its relative molecular mass? [1]

    (c) State whether or not it reacts with each of the following and, where it does react, write the formula of the product:

        (i) Tollens' reagent   **(ii)** Fehling's solution

        (iii) PCl_5 **(iv)** $KMnO_4/H^+$. [4]

32.6* Complete the analysis of functional groups below by listing the groups corresponding to **(i)–(xi)**.

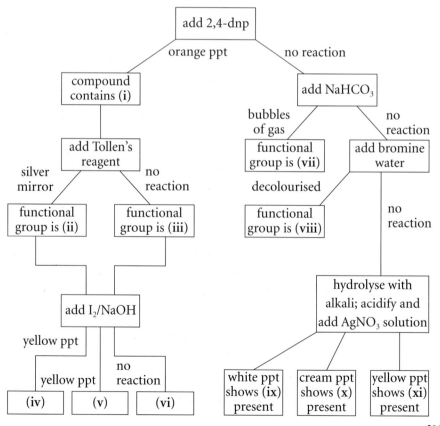

[11]

- the uses of ultra-violet and visible spectroscopy in the elucidation of structure
- the interpretation of infra-red spectra and the use of characteristic frequencies to identify some functional groups
- the use of nuclear magnetic resonance spectroscopy to determine the location and number of protons in organic molecules
- the interpretation of spin-spin coupling to obtain structural information
- the use of mass spectra in determining relative molecular masses
- the interpretation of fragmentation patterns to provide structural information
- the combined use of several spectroscopic techniques in the determination of structure.

33.1 This question is concerned with nuclear magnetic resonance (NMR) spectroscopy.
  (a) (i) Why is an 'internal standard' required in NMR spectroscopy?
    (ii) What substance is commonly used as an internal standard in proton NMR spectroscopy?
    (iii) Explain why this is a particularly good internal standard for proton NMR spectroscopy.
    (iv) What chemical shift is assigned to the protons in this internal standard?
  (b) (i) Give **two** reasons for the widespread use of tetrachloromethane as a solvent in NMR spectroscopy.
    (ii) What solvent is employed to obtain NMR spectra of substances which are significantly soluble only in water? **[11]**

33.2 (a) Suggest how the isomers $CH_3—O—CH_3$ and C_2H_5OH might be distinguished using NMR and infra-red (IR) spectroscopy. **[5]**
  (b) How might their mass spectra provide further evidence to distinguish them? **[1]**

33.3 A student is attempting to reduce propanone, CH_3COCH_3, using sodium borohydride. How might the student determine whether the reduction was complete using infra-red spectroscopy? **[2]**

33.4 (a) (i) Sketch the expected proton NMR spectrum of 1,1,2-trichloroethane, $CHCl_2—CH_2Cl$.
    (The chemical shift of the CH_2 $Cl—$ protons is 4.0δ/ppm and of the $—CHCl_2$ proton is 6.0δ/ppm.)

(ii) How does the spectrum confirm that the hydrogens in the molecule are on adjacent carbon atoms? [5]

(b) (i) Sketch the expected mass spectrum of 1,1,2-trichloroethane in the region of its parent ion.

(ii) How might mass spectroscopy enable it to be distinguished from its structural isomer 1,1,1-trichloroethane, CCl_3—CH_3? [5]

33.5 A substance X having the molecular formula $C_4H_8O_2$ has strong infra-red absorptions at 1750 and 1250 cm^{-1}. The mass and proton NMR spectra of X are shown below. Work out the structural formula of X and name the class of compounds to which it belongs. [10]

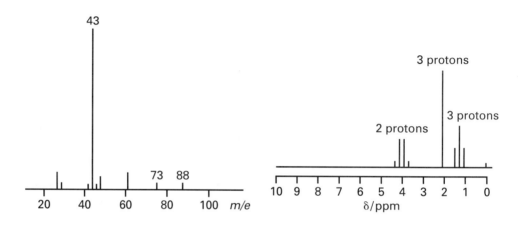

33.6* The structures of two isomers having the molecular formula C_7H_7Cl are shown below.

1-chloro-4-methylbenzene   (chloromethyl)benzene

State and explain which of infra-red, nuclear magnetic resonance or mass spectrometry would be:

(a) least useful in distinguishing the two isomers

(b) most useful in distinguishing the two isomers. [5]

33.7 The wavelengths of various colours in visible light are shown in the table.

colour	violet	blue	green	yellow	red
wavelength/nm	400	470	500	600	650

Use the information to suggest what colour the compounds whose visible spectra appear below would be. **[6]**

(a)

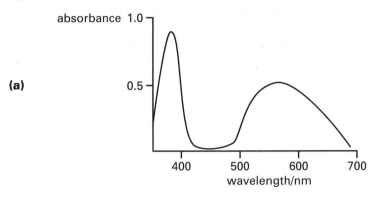

(b)

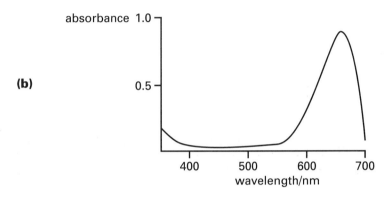

(c)

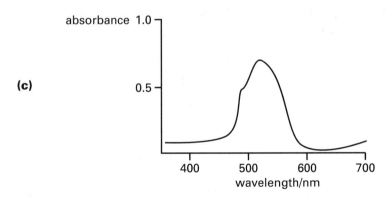

33.8 State the splitting patterns appropriate to the hydrogen atom(s) indicated in the following molecules:

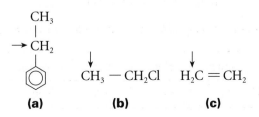

(a)   **(b)**   **(c)**

33.9* Compound X is a colourless liquid with a boiling point of 100 °C. It contains 91.3% carbon and 8.7% hydrogen. Details of the infra-red, mass and proton NMR spectra of X appear below.

mass spectrum	peaks at m/e 92,91 (both of high abundance but few other peaks of diagnostic use)
IR spectrum	absorptions at 3044, 2938, 1600, 1494, 720 (strong) cm^{-1}
NMR spectrum	peaks at 7.0δ (singlet, 5 protons) 2.2δ (singlet, 3 protons)/ppm

(a) Identify X, giving your reasoning. **[8]**

(b) X can be oxidised with acidified potassium manganate(VII). Name the compound which would result and suggest at what frequencies significant additional absorptions might be expected in its infra-red spectrum. **[3]**

(c) Suggest how the existence of a strong peak at m/e 91 and the absence of a peak at m/e 15 might be linked. **[3]**

33.10 A substance, A, has the following percentage composition by mass: C 30.6, H 3.8, Cl 45.2 and O 20.4. Substance A reacts vigorously with water at room temperature. The main features of its mass, infra-red and NMR spectra are given below:

mass spectrum	peaks at m/e, 80, 78, 43, 35, 37, 15
IR spectrum	strong absorptions at 1800 and 600 cm^{-1}
NMR spectrum	single peak at 1.8δ/ppm

Suggest a structural formula for A, explaining your reasoning. **[9]**

33.11* A compound B contains 66.7% carbon, 11.1% hydrogen and 22.2% oxygen. Its infra-red spectrum shows a strong absorption at 1700 cm^{-1} and significant peaks appear in its mass spectrum at m/e values 72, 57, 43 and 29. Details of its NMR spectrum are summarized below:

chemical shift (δ/ppm)	multiplicity	relative intensity
2.5	quartet	2
2.1	singlet	3
1.0	triplet	3

(a) Suggest a structure for B, explaining your reasoning. **[14]**

(b) Give a chemical test, and the expected result, which would support your answer to part **(a)** **[5]**

(c) **(i)** Write down the structural formula and name of an isomer of B.

 (ii) Give a chemical test, and the expected result, which would enable this isomer to be distinguished from B. **[4]**

33.12*   Spectral data on a compound C are summarized below:

mass spectrum	peaks at m/e 58, 57, 29, 28, 27
IR spectrum	absorptions at 3000, 2800, 2700 and 1730 cm^{-1}
NMR spectrum	9.8δ (singlet, 1 proton), 2.5δ (quartet*, 2 protons), 1.1δ (triplet, 3 protons)/ppm *At higher resolution each line in this quartet appears as a doublet.

Suggest a structure for compound C on the basis of this information.   **[9]**

33.13*   **(a)** Use the information below to identify compound A.
Elemental analysis (% by mass): C 49.3, H 9.7, N 19.0.
A reacts with phosphorus(V) oxide to give a compound of molecular formula C_3H_5N and ammonia is evolved on heating A with aqueous alkali.

mass spectrum	peaks at m/e 73, 57, 44, 29
IR spectrum	strong absorptions at 3340, 3200 and 1640 cm^{-1}
NMR spectrum	6.2δ (broad line, 2 protons), 2.2δ (quartet, 2 protons), 1.1δ (triplet, 3 protons)/ppm **[10]**

     (b) Give the reagents and conditions required to convert A to ethylamine.   **[3]**

     (c) Explain why compound A is a much weaker base in aqueous solution than ethylamine.   **[3]**

33.14*   Use the information below to identify compound B.
Elemental analysis (% by mass): C 28.74, H 4.19, O 19.16, Br 47.91.

mass spectrum	peaks at m/e 168, 166 and a very intense peak at m/e 87
IR spectrum	strong absorptions at 1740 and 1200 cm^{-1}
NMR spectrum	3.7δ (singlet, 3 protons), 3.5δ (triplet, 2 protons), 2.9δ (triplet, 2 protons)/ppm **[10]**

33.15*   **(a)** Use the information below to identify compound C.
C contains carbon and hydrogen only. Oxidation with alkaline potassium manganate(VII) gives benzoic acid.

mass spectrum	peaks at m/e 106 (parent ion) and a very intense peak at m/e 91
NMR spectrum	7.1δ (singlet, 5 protons), 2.7δ (quartet, 2 protons), 1.2δ (triplet, 3 protons)/ppm **[6]**

     (b) Give the reagents and conditions required to prepare C from benzene.   **[4]**

     (c) Suggest a formula for the very intense peak at m/e 91 in the mass spectrum and suggest how it might have been formed in the mass spectrometer.   **[3]**

33.16* Use the information below to identify compound E, which has the molecular formula $C_6H_{10}O_3$. The infra-red spectrum shows evidence for an ester linkage and a ketonic carbonyl.

mass spectrum	peaks at m/e 130, 115, 87, 85 and 43
NMR spectrum	4.1δ (quartet, 2 protons), 3.3δ (singlet, 2 protons), 2.2δ (singlet, 3 protons) and 1.2 (triplet, 3 protons)/ppm **[8]**

Part 3

Synoptic questions

The questions in this section are labelled according to the nature of the chemistry involved: **P** = physical, **O** = organic, **I** = inorganic, **A** = analytical.

1 Explain each of the following observations as fully as you can in terms of the structure of the substances involved and the types of bond present.

(a) The boiling point of silicon(IV) oxide is 2230 °C, while that of carbon(IV) oxide (carbon dioxide) is −78 °C. **[4]**

P

(b) Metals are good electrical conductors and are ductile. Copper is a better electrical conductor than potassium and lithium is harder than caesium. **[4]**

(c) Sodium chloride is readily soluble in water, but insoluble in tetrachloromethane, whereas the reverse is true of iodine. **[7]**

(d) The boiling points (in °C) of the hydrides of the first three elements of group 5 are as follows: NH_3 −33, PH_3 −88, AsH_3 −57. **[4]**

(e) Diamond is hard and an electrical insulator, while graphite is soft and conducts electricity well. **[4]**

2 **(a)** The solid sodium halides give different products when they react with concentrated sulphuric acid.

(i) Copy and complete the table below for NaBr and NaI.

I/A

halide	product containing a halogen	product containing sulphur
NaCl	HCl	$NaHSO_4$
NaBr		
NaI		**[4]**

(ii) Explain the differing reaction observed for NaI and NaCl. **[4]**

(b) Copy and complete the table below to show how the halide ions Cl^-, Br^- and I^- may be distinguished in aqueous solution by the addition of aqueous silver(I) ions, followed by aqueous ammonia.

halide ion	observation with aqueous silver(I) ions	observation when aqueous ammonia is added to halide ion plus silver(I) ions
Cl^-		
Br^-		
I^-		**[6]**

(c) Chlorate(I) ions disproportionate on heating in aqueous solution.

(i) Write an equation for the disproportionation reaction.

(ii) Use your equation to explain the meaning of the term 'disproportionation'. [4]

(d) Explain the following observations:

(i) chlorine is a more powerful oxidizing agent than bromine

(ii) aqueous solutions of hydrogen fluoride are only weakly acidic. [8]

3 The table below shows some standard electrode potentials.

reaction	$E^{\ominus}/V$
$Fe^{2+}(aq) + 2e^- \rightleftharpoons Fe(s)$	-0.44
$Cu^{2+}(aq) + 2e^- \rightleftharpoons Cu(s)$	$+0.34$
$I_2(aq) + 2e^- \rightleftharpoons 2I^-(aq)$	$+0.54$
$Fe^{3+}(aq) + e^- \rightleftharpoons Fe^{2+}(aq)$	$+0.77$
$Br_2(aq) + 2e^- \rightleftharpoons 2Br^-(aq)$	$+1.07$
$Cl_2(aq) + 2e^- \rightleftharpoons 2Cl^-(aq)$	$+1.36$

P

(a) Standard electrode potentials can be measured by reference to the standard hydrogen electrode.

(i) Give the $E^{\ominus}$ value of the standard hydrogen electrode.

(ii) Give the standard conditions in the standard hydrogen electrode when standard electrode potentials are being measured. [4]

(b) Give the formula of the species in the table which, under standard conditions, is:

(i) the most powerful reducing agent

(ii) the most powerful oxidizing agent.

(iii) Which of the halogens would oxidize $Fe^{2+}(aq)$ to $Fe^{3+}(aq)$ under standard conditions? [4]

(c) (i) What is the sign of the electrode potential of a spontaneous redox reaction?

(ii) In the laboratory, reactions which should occur on the basis of the standard electrode potentials of the reagents sometimes fail to do so. Suggest **two** reasons for this. [3]

(d) An electrochemical cell is set up involving iron and copper dipping into molar solutions of their respective +2 ions.

(i) Write down the conventional representation of this cell.

(ii) Calculate the e.m.f. of the cell. [4]

(e) (i) At which metal electrode will reduction occur?

(ii) Write down the equation for the reaction at this electrode.

(iii) Write an equation to represent the overall cell reaction when the cell is used to supply a current. [5]

4 An oxide of an element of Group 1, M_2O, contains 83.0% of the metal. On further heating in oxygen M_2O reacts to give another oxide Y, which contains 71.0% of the metal.

(a) (i) Calculate the relative atomic mass of M and hence identify it.

(ii) Write down the electronic configuration of M and use it to explain why M forms only a $+1$ ion. [7]

I

(b) Write an equation to represent the first ionization energy of M. [2]

(c) (i) Describe the reaction of M with water and give an equation for the reaction.

(ii) M reacts less readily with water than the element below it in Group 1. Give an explanation for this. [9]

(d) (i) Calculate the empirical formula of the oxide Y and write down the formula of the ions which it contains.

(ii) Suggest why the oxide of lithium corresponding to Y is unstable compared to those of the other Group 1 metals. [5]

5 (a) Magnesium has three isotopes, having the following percentage abundances

P/A

mass number	atomic number	% abundance
24	12	78.6
25	12	10.1
26	12	11.3

(i) Give the number of protons, neutrons and electrons in the isotope of mass number 25.

(ii) Calculate the relative atomic mass of magnesium.

(iii) Sketch the expected mass spectrum of magnesium. [7]

(b) (i) Write an equation representing the first ionization energy of an element M.

(ii) Explain why ionization energies decrease as atomic number increases within the group.

(iii) Explain why ionization energies generally increase with atomic number across a period. [6]

(c) Explain why:

(i) the first ionization energy of boron is less than that of beryllium

(ii) the first ionization energy of oxygen is less than that of nitrogen. [4]

6 This question relates to the Group 2 elements Mg, Ca, Sr and Ba and their compounds.

I

(a) Write down the symbol for the element which:

(i) forms the most soluble hydroxide

(ii) forms the least soluble sulphate

(iii) forms the smallest $+2$ cation

(iv) reacts most vigorously with water. [4]

(b) State and explain the effect of increasing cation size on the thermal stability of the carbonates of Group 2. **[5]**

(c) Group 2 metal nitrates decompose on heating according to the equation:

$$2M(NO_3)_2(s) \rightarrow 2MO(s) + 4NO_2(g) + O_2(g)$$

0.424 g of a nitrate gave 120 cm³ of gas on complete decomposition. (All measurements were made at r.t.p. The molar volume of a gas at r.t.p. is approximately 24 dm³.)

(i) Identify the metal M, showing your working clearly.

(ii) What would be the mass of the solid remaining after the reaction? **[6]**

7 (a) Give the formula of an oxide of the following elements, together with the acid–base character of the oxide: sodium, sulphur, aluminium. **[3]**

P/I

(b) (i) From the oxides of the elements Na to Ar, name one which has: an ionic lattice, a simple molecular structure, a giant covalent structure.

(ii) Give **one** difference and **one** similarity between the properties of substances having ionic lattices and substances having giant covalent structures and explain each. **[9]**

8 Sodium hydroxide is manufactured by electrolysing saturated sodium chloride solution (brine) in a diaphragm cell, shown in the diagram.

I

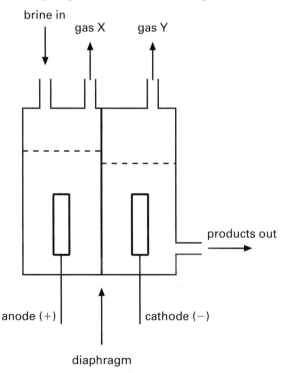

(a) Give balanced equations for the formation of the gases X and Y. **[2]**

(b) (i) In which compartment, anode or cathode, is sodium hydroxide formed?

(ii) Explain why sodium hydroxide forms in this compartment. **[5]**

(c) If the diaphragm is removed an anion, Z, containing chlorine is formed. The anion, Z, disproportionates if an aqueous solution of it is boiled.

  (i) Write a balanced ionic equation for the formation of Z.

  (ii) Write a balanced ionic equation for the disproportionation of Z and use it to explain the meaning of the term 'disproportionation'. **[6]**

(d) (i) State and explain the conditions under which sodium chloride must be electrolysed to manufacture sodium.

  (ii) Give **one** commercial use each for sodium and chlorine. **[2]**

9 Nitrite ions can be determined quantitatively by titration with manganate(VII) ions in acidic solution, according to the equation:

$$2MnO_4^- + 5NO_2^- + 6H^+ \rightarrow 2Mn^{2+} + 3H_2O + 5NO_3^-$$

(a) Iron(II) ammonium sulphate ($FeSO_4.(NH_4)_2SO_4.6H_2O$, molar mass $= 392$ g mol^{-1}) can be used to standardize the solution of manganate(VII) ions.

  (i) Write an ionic equation for the reduction of manganate(VII) ions to Mn^{2+} by iron(II) in acidic solution.

  (ii) What mass of iron(II) ammonium sulphate would be used to prepare 250 cm^3 of a solution of concentration 0.1 mol dm^{-3}? **[5]**

P

(b) Write the two half equations for the overall reaction between manganate(VII) ions and nitrite ions in acidic solution. **[4]**

(c) In a typical experiment to determine the concentration of nitrite ions, 25.0 cm^3 of a 0.02 M solution of potassium manganate(VII) was acidified, heated to about $40\,°C$ and then titrated with a solution of sodium nitrite, of which 25.0 cm^3 were required to reach the end-point.

  (i) How might the end-point of the titration be most conveniently determined?

  (ii) Calculate the concentration, in mol dm^{-3}, of nitrite ions in the solution. **[3]**

(d) The aqueous Mn^{3+} ion is as powerful an oxidizing agent as manganate(VII), but it is rarely used because it readily disproportionates into solid MnO_2 and Mn^{2+} ions.

  (i) What is meant by 'disproportionates'?

  (ii) Write a balanced equation for the disproportionation of the Mn^{3+} ion into MnO_2 and Mn^{2+}.

  (iii) State and explain how the tendency of the Mn^{3+} ion to disproportionate would be affected by changes in the pH of the reaction mixture. **[7]**

10 (a) The elements in the third period of the Periodic Table are each represented by the symbol E. For each of the following, what is the correct symbol for an element which forms:

  (i) an oxide having formula E_2O_3 only

  (ii) oxides having formulae E_2O and E_2O_7

  (iii) an oxide of formula EO_2 only? **[4]**

I

(b) For each of the following, name one oxide in the third period which has:
  (i) an ionic lattice
  (ii) a giant molecular structure
  (iii) a simple molecular structure. [3]
(c) Give **one** similarity and **one** difference between substances having ionic lattices and giant molecular structures. [2]

11 (a) Define the first ionization energy of an element. [1]
(b) Write equations, including state symbols, to show the changes involved in:
  (i) the first ionization energy of magnesium [1]
  (ii) the second ionization energy of magnesium. [1]
(c) Explain why the second ionization energy of magnesium ($+1450$ kJ mol^{-1}) is greater than the first ionization energy ($+736$ kJ mol^{-1}). [3]
(d) The first ionization energies (in kJ mol^{-1}) of the elements of period 2 of the Periodic Table are shown in the table below.

Li	Be	B	C	N	O	F	Ne
519	900	799	1090	1400	1310	1680	2080

Explain each of the following observations.
  (i) There is a general increase in first ionization energy from Li to Ne. [3]
  (ii) The first ionization energy of oxygen is lower than that of nitrogen. [4]
(e) Naturally-occurring copper consists of two isotopes having the following atomic masses and percentage abundance: ^{63}Cu 69.1% and ^{65}Cu 30.9%. Calculate the relative atomic mass of naturally-occurring copper. [2]
(f) A newly-discovered compound is thought to be the covalent substance copper dimethyl, $Cu(CH_3)_2$. With reference to its mass spectrum, give:
  (i) the **two** highest expected m/e values
  (ii) the formula and m/e value of **two** other possible ions both of which contain carbon which might be observed. [4]

12 State and explain the differences in the physical properties given for each of the following pairs of substances, in terms of their bonding and structure.
(a) The melting points of ethene and poly(ethene).
(b) The electrical conductivities of graphite and diamond.
(c) The shapes of boron trifluoride (BF_3) and ammonia (NH_3).
(d) The ductilities of sodium and sulphur. [4 × 4]

13 The mass spectrum of an element enables the relative abundance of each isotope of the element to be determined. Data relating to the mass spectrum of bromine appear on the following page.

atomic number	mass number of isotope	% relative abundance
35	79	50.5
	81	49.5

A

(a) Define the term 'isotope'. [2]

(b) Write down the conventional symbols for the two isotopes of bromine. [2]

(c) Calculate the precise relative atomic mass of bromine to four significant figures. [2]

(d) (i) Sketch the expected mass spectrum of bromine vapour in the m/e region 158 to 162. [5]

(ii) Explain the origin of each peak in your spectrum. [3]

14 This question concerns the chemistry of some of the Group 2 elements listed below.

element	atomic number
magnesium	12
calcium	20
strontium	38
barium	56

I

(a) Name the element which forms:

(i) the most soluble sulphate

(ii) the least soluble hydroxide

(iii) the most thermally stable nitrate

(iv) compounds which impart a green colour to a Bunsen flame

(v) compounds which impart no colour to a Bunsen flame. [5 × 1]

(b) (i) Describe and explain the changes in atomic radius which occur with increase in atomic number as the group is descended.

(ii) State and explain the relative sizes of the Na^+ and Mg^{2+} ions. [2 × 3]

(c) Strontium may be prepared by electrolysing fused strontium chloride. The molten metal collects at the cathode and must be protected from air and moisture.

(i) Name the product formed at the anode.

(ii) Name a suitable anode material, giving a reason for your choice.

(iii) Explain briefly why chemical reduction cannot be used to extract strontium from its chloride. [5]

(d) (i) Name and give the formula of the substance which would form if air reached the molten strontium.

(ii) Write a balanced equation for the reaction of this substance with water. [4]

15* Chlorine disproportionates in water at room temperature according to the
following equation:

$$Cl_2(g) + H_2O(l) \rightleftharpoons HClO(aq) + HCl(aq)$$

(a) (i) Give the names and oxidation numbers of the anions formed in this
reaction.

(ii) What does the term 'disproportionates' mean?

(iii) Suggest why the acidity of the reaction mixture cannot be determined
directly by titration with alkali. **[7]**

(b) On boiling the solution, one of the ions undergoes further
disproportionation.

(i) Which one of the ions disproportionates and why can it alone do so?

(ii) Write a balanced equation for this disproportionation reaction. **[5]**

(c) When fluorine is bubbled into water at room temperature, oxygen is evolved
and a weakly-acidic solution results. The reaction below occurs:

$$2F_2(g) + 2H_2O(l) \rightarrow 4HF(aq) + O_2(g)$$

(i) Suggest **two** reasons why oxygen is produced with fluorine but not if
chlorine is used instead of fluorine.

(ii) Give **two** factors contributing to the fact that fluorine fails to form stable
oxoanions analogous to those formed by chlorine.

(iii) Why is the resulting solution **weakly** acidic? **[7]**

(d) Gaseous hydrogen fluoride may be prepared in the laboratory by heating
solid sodium fluoride with concentrated sulphuric acid.

(i) Name the gaseous products formed when the reaction is repeated using
sodium bromide.

(ii) Explain any differences between the two reactions. **[6]**

16 Use the simplified Periodic Table below to answer the questions which follow.

1	2											3	4	5	6	7	0
Li	Be											B	C	N	O	F	Ne
Na	Mg											Al	Si	P	S	Cl	Ar
K	Ca	Sc	Ti	V	Cr	Mn	Fe	Co	Ni	Cu	Zn	Ga	Ge	As	Se	Br	Kr
Rb	Sr												Sn			I	Xe
Cs	Ba												Pb				

(a) From the s-block elements:

(i) Give the symbol for the element which reacts most exothermically with
water. Give an equation for this reaction. Suggest why this element reacts
most exothermically with water.

    (ii) Give the symbol for the element which forms the most insoluble
hydroxide. How does the solubility of the hydroxides in group 2 vary
with the position of an element in the group? **[7]**

(b) From the elements in the d-block:

    (i) Give the formula of the ion of an element in the $+2$ oxidation
state which contains six d-electrons, four of them being unpaired.
Suggest why this ion is relatively easy to oxidize to the $+3$ oxidation
state.

    (ii) Give the symbol for the element forming only one oxidation state.
Explain why only this single oxidation state is known.

    (iii) Give the formula of an ion having the highest oxidation state for any of
the elements. Give the oxidation state of the metal in this ion. **[8]**

(c) From the p-block elements:

    (i) Give the symbol for the element having the highest boiling point in
Group 7. Give the formula for the molecules of this element. Explain
why this element has the highest boiling point in Group 7.

    (ii) Give the formula of an oxide of the element showing most metallic
properties. Give a reason for choosing this element as the most
metallic.

    (iii) Give the symbol of an element forming an acidic hydride. Explain why
this hydride is acidic and write an equation for its reaction with aqueous
sodium hydroxide solution. **[15]**

17 Copy and complete the table below.

		number of bond pairs	number of lone pairs	shape of molecule	formula of a molecule having the shape given
P	**(a)**	4	0		
	(b)	3	1		
	(c)	2	2		
	(d)	1	3		**[4 × 2]**

18 This question is about Group 7 of the Periodic Table.

(a) State the colour and physical state at room temperature of fluorine, chlorine,
bromine and iodine. **[4]**

(b) When potassium chloride is mixed with concentrated sulphuric acid and
gently warmed, a steamy gas, Y, is evolved.

    (i) Write an equation for this reaction.

    (ii) When Y is dissolved in water, in which it is readily soluble, a strongly
acidic solution, Z, is formed. Describe the reaction that occurs between
Y and water in terms of the Brønsted–Lowry theory.

(iii) Silver nitrate solution is added to Z. A white precipitate forms. If silver nitrate in concentrated aqueous ammonia is cautiously added to Z, no precipitate forms. Explain these observations. Why must the second addition be carried out cautiously? [10]

(c) Aqueous chlorate(I) ions, ClO^-, decompose on warming in a disproportionation reaction. Write the equation for this reaction, and by considering the oxidation state changes involved, explain the term 'disproportionation'. [3]

(d) Chlorine is extensively used in water treatment plants.

(i) Upon which chemical and physical properties of chlorine does its use rely?

(ii) Excess chlorine can be removed by treatment with aqueous sulphur dioxide, which is oxidized to sulphate ion. Write a balanced equation for the reaction between chlorine and sulphur dioxide. [5]

19 (a) 1.160 g of a Group 1 element, M, reacts with oxygen to give 1.398 g of an oxide, M_2O. 1.00 mole of M_2O reacts further with 0.50 moles of oxygen gas on heating to give a compound, X, having the empirical formula MO.

(i) Identify the metal M.

(ii) Write a balanced equation, with state symbols, for the reaction of M_2O with oxygen.

(iii) Give the formula of the anion present in compound X.

(iv) What is the oxidation state (number) of oxygen in this anion?

(v) Explain why this anion might be expected to behave as an oxidizing agent. [8]

(b) Radium (atomic number 88) is the last member of Group 2. State which of the following alternatives is the correct response and give an explanation of your choice of answer.

(i) Radium would be harder/softer than magnesium.

(ii) The Ra^{2+} ion would have a larger/smaller radius than Ba^{2+}.

(iii) Radium metal would react more/less vigorously with water than calcium.

(iv) Radium carbonate would decompose/not decompose into the corresponding oxide and carbon dioxide at a lower temperature than calcium carbonate.

(v) Radium sulphate would/would not dissolve to a significant extent in water at room temperature. [14]

20 This question is concerned with the hydrogen halides, HX (X = F, Cl, Br and I).

(a) (i) Name the hydrogen halide which behaves as a weak acid in aqueous solution.

(ii) Explain the meaning of the term 'weak' in this context.

(iii) Why does the hydrogen halide you have named behave as a weak acid? [4]

A/I

(b) (i) Name the hydrogen halide having the highest boiling point.

(ii) Explain why it has the highest boiling point. **[4]**

(c) (i) Name the principal intermolecular force between molecules of HBr and HI.

(ii) Which of these has the higher boiling point, and why? **[3]**

(d) Concentrated sulphuric acid reacts with sodium halides to form products containing the corresponding halogen.

(i) Name a sodium halide which would form a hydrogen halide when treated in this way.

(ii) Write a balanced equation for the reaction. **[3]**

(e) Some sodium halides react further with concentrated sulphuric acid. **[3]**

(i) Write a balanced equation for **one** such reaction.

(ii) Explain, in terms of the oxidation state changes which take place, what has been oxidized and what has been reduced in the reaction.

(iii) How do you account for the further reaction with sulphuric acid of the sodium halide which you have chosen? **[9]**

21 Use the following standard electrode potentials to answer the questions which follow.

P/I

$$V^{3+}(aq) + e^- \rightleftharpoons V^{2+}(aq) \qquad\qquad E^\ominus = -0.26 \text{ V}$$
$$VO^{2+}(aq) + 2H^+(aq) + e^- \rightleftharpoons V^{3+}(aq) + H_2O(l) \qquad E^\ominus = +0.34 \text{ V}$$
$$VO_2^+(aq) + 2H^+(aq) + e^- \rightleftharpoons VO^{2+}(aq) + H_2O(l) \qquad E^\ominus = +1.00 \text{ V}$$
$$SO_4^{2-}(aq) + 2H^+(aq) + 2e^- \rightleftharpoons SO_3^{2-}(aq) + H_2O(l) \quad E^\ominus = +0.17 \text{ V}$$

(a) (i) Explain which oxidation state of vanadium would be formed when sulphite ions (SO_3^{2-}) are added to vanadium(V) in aqueous solution.

(ii) Write a balanced redox equation for the reaction which occurs.

(iii) What colour would the resulting solution be? **[7]**

(b) (i) What is meant by the term 'disproportionation'?

(ii) Write a balanced equation for the disproportionation of V(IV) to V(III) and V(V) and predict, giving a reason, whether it would occur under standard conditions.

(iii) Explain what effect an increase in the pH would have on the tendency of V(IV) to disproportionate. **[11]**

(c) Vanadium forms a complex having the formula $[VCl_2(H_2O)_4]Cl$.

(i) What is the oxidation state of vanadium in this complex?

(ii) Name the complex. **[3]**

22 (a) (i) Give the meaning of the terms 'd-block element' and 'transition metal'.

(ii) Name **two** elements in the d-block which are *not* transition metals.

(iii) Give **two** characteristic properties of transition metals. **[7]**

(b) (i) Write down the electronic configuration of Cu, Cu^+ and Cu^{2+}.

(ii) What colour would you expect aqueous solutions of Cu^+ to have? Explain your answer. **[6]**

(c) The highest oxidation state of iron can be isolated as a barium salt, $BaFeO_n$, where n represents the number of oxygens. The anion in the salt, FeO_n^{2-}, is a powerful oxidizing agent, being reduced in the process to Fe^{3+} ions. On treatment with excess potassium iodide, 1 mole of FeO_n^{2-} liberates 1.5 moles of iodine.

(i) Calculate the oxidation number of iron in FeO_n^{2-}.

(ii) What is the value of n?

(iii) Using your value of n from (ii), write down a balanced redox equation for the reaction between the anion and iodide ions in aqueous solution.

(iv) What shape would you expect the anion to be?

(v) Write down the electronic configuration of iron in the anion. [7]

23 Explain the following observations, giving equations where relevant.

(a) The addition of aqueous sodium hydroxide to copper(II) sulphate solution forms a blue precipitate which is insoluble in excess sodium hydroxide, but dissolves in excess aqueous ammonia to give a deep-blue solution. [7]

(b) Iron wool glows when heated in dry chlorine, forming a blackish-green product. It behaves similarly when heated in hydrogen chloride gas, but the product is white. In the presence of water the products from both reactions have colours different from those obtained when no water is present. [7]

(c) Iron is protected from corrosion when placed in contact with a piece of zinc, but it corrodes more rapidly when placed in contact with a piece of tin. You should base your answer on the following standard electrode potentials: Zn^{2+}/Zn -0.76 V; Fe^{2+}/Fe -0.44 V; Sn^{2+}/Sn -0.14 V. [4]

(d) Three chromium(III) complexes A, B and C having the molecular formula $CrCl_3(H_2O)_6$ can be prepared. The results of adding aqueous silver nitrate to their aqueous solutions are described below. There are two possible structures for the cation in complex A.

complex	number of moles of AgCl immediately precipitated per mole of complex
A	1
B	2
C	3

[6]

24 The following question concerns the transition elements. Explain the following, giving equations where relevant.

(a) The number of stable oxidation states shown by the d-block elements increases from scandium to manganese but decreases markedly thereafter. [5]

(b) Scandium and zinc, though found in the d-block in the Periodic Table, are not classed as transition elements. [4]

(c) Manganese and iron both react readily with acids to form solutions of their respective +2 ions but manganese is the more powerful reducing agent. **[5]**

(d) Iron gives different products when heated with chlorine and with hydrogen chloride. Aqueous solutions of the product formed with hydrogen chloride are easily oxidized to that formed with chlorine. **[6]**

25 **(a)** The reaction which occurs in the Leclanché (dry) cell is represented by the equation

$$Zn(s) + 2NH_4^+(aq) + 2MnO_2(s) \rightarrow Zn^{2+}(aq) + 2NH_3(aq) + H_2O(l) + Mn_2O_3(s).$$

(i) For each of the elements in the following species, give the change in oxidation state which occurs during this reaction:

Zn, MnO_2, NH_4^+.

(ii) Which of these species undergoes reduction as the cell reaction proceeds? **[4]**

P/I **(b) (i)** Write down and balance the equations for the half-reactions in which: aqueous $Mn(III)$ ions are reduced to aqueous $Mn(II)$ ions; solid MnO_2 is reduced to aqueous $Mn(III)$ ions.

(ii) Define the term 'disproportionation'.

(iii) Write a balanced equation for the disproportionation of $Mn(III)$ ions into $Mn(II)$ ions and solid MnO_2.

(iv) State and explain how the stability of $Mn(III)$ ions to disproportionation would be affected by a decrease in pH. **[12]**

(c) Consider the following standard electrode potentials:
$$E^{\ominus}Mn^{3+}/Mn^{2+} = +1.51 \text{ V}, E^{\ominus}Fe^{3+}/Fe^{2+} = +0.77 \text{ V}.$$

(i) Which +3 ion is the stronger oxidizing agent?

(ii) Explain your answer to **(i)** in terms of the electronic configurations of the ions involved. **[4]**

26 **(a)** What is meant by the term 'transition element'? **[2]**

(b) Write down the electronic structures of the following ions:
Fe^{2+}, Fe^{3+}, Cu^+, Cu^{2+}. **[4]**

(c) Give **three** characteristic properties of transition elements and/or their compounds. **[3]**

(d) For *either* iron *or* copper, give an example to illustrate each of these characteristic properties. **[3]**

27 The manganate(VII) ion, MnO_4^-, is widely used in the laboratory as an oxidizing agent. When it behaves as an oxidizing agent it is converted to the Mn^{2+} ion.

(a) Give the oxidation number of manganese in MnO_4^- and Mn^{2+}. **[2]**

(b) Is the conversion of MnO_4^- to Mn^{2+} an oxidation or a reduction? Explain your answer. **[2]**

28 This question concerns the d-block elements from scandium to zinc.

(a) (i) Write down the electronic configuration of each of the following ions:

		3d					4s
Cu$^+$	[Ar]						
Cr^{2+}	[Ar]						
Sc^{3+}	[Ar]						

(ii) Which of the above ions would form colourless aqueous solutions?

(iii) What feature of the electronic configuration of the ions identified in part (ii) is responsible for their lack of colour? **[7]**

(b) Cobalt forms a compound of formula $[Co(NH_3)_4Cl_2]^+Cl^-$.

(i) What is the oxidation state and electronic configuration of cobalt in this compound?

(ii) Give the name of the complex ion in the compound.

(iii) What type of bond exists between the cobalt ion and its ligands?

(iv) State and explain how many moles of silver chloride would be immediately precipitated from one mole of this compound in aqueous solution by the addition of an excess of aqueous silver nitrate. **[9]**

29 The first row d-block elements are scandium to zinc, inclusive.

(a) Name **two** of these elements which are not considered to be transition elements, giving a different reason for their exclusion in each case. **[4]**

(b) Transition metal ions form complexes with ligands, usually containing four or six ligands, for example $Ni(CN)_4^{2-}$ and $Fe(CN)_6^{3-}$.

(i) What is the oxidation state (number) of nickel in $Ni(CN)_4^{2-}$?

(ii) Write down the formula of (A) another charged ligand (excluding CN^-) and (B) a neutral ligand.

(iii) Name a possible shape for $Ni(CN)_4^{2-}$.

(iv) What type of bond exists between the transition metal ion and the ligands in such complexes? **[6]**

(c) When an aqueous solution of copper(II) sulphate is treated dropwise with aqueous ammonia solution a blue precipitate forms, which dissolves to give a deep-blue solution when excess of aqueous ammonia solution is added. Explain, giving equations:

(i) the formation of the blue precipitate.

(ii) the dissolving of the blue precipitate in aqueous ammonia solution.

(iii) the deep-blue colour of the solution obtained when excess aqueous ammonia solution is added, compared with the pale blue of the aqueous copper(II) sulphate solution. **[12]**

30 The apparatus shown on the following page was used to measure the standard electrode potential of the $M^{2+}(aq)/M(s)$ electrode.

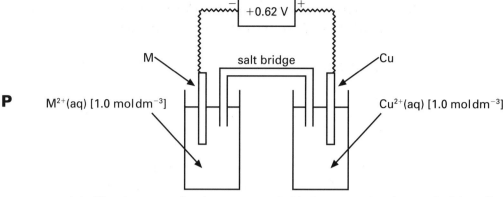

(a) Give the name of an instrument suitable for measuring the e.m.f. of the cell and indicate its main characteristic. **[2]**

(b) (i) Suggest how the salt bridge might be prepared.

(ii) What is the purpose of the salt bridge? **[3]**

(c) The e.m.f. of the cell is 0.62 V. If the standard electrode potential of the Cu^{2+}/Cu electrode is +0.34 V, what is the standard electrode potential of the M^{2+}/M electrode? **[2]**

(d) (i) Write an equation for the reaction occurring in the cell when a current is drawn, explaining your reasoning.

(ii) Write down the conventional representation of the cell.

(iii) State and explain the effect on the cell e.m.f. of diluting the aqueous solution of Cu^{2+}. **[9]**

31 Chromium(III) chloride, $CrCl_3$, reacts with water to form a hydrate, $CrCl_3.6H_2O$, which exists as three structural isomers represented by the following formulae:

$$A\ [Cr(H_2O)_6]^{3+}3Cl^- \quad \text{(violet)}$$
$$B\ [Cr(H_2O)_5Cl]^{2+}2Cl^-.H_2O \quad \text{(pale green)}$$
$$C\ [Cr(H_2O)_4Cl_2]^+Cl^-.2H_2O \quad \text{(dark green)}$$

(a) Give **two** characteristic properties of transition elements shown by chromium in these isomers. **[2]**

(b) (i) Describe a test, and the expected result, which would show that each isomer contained free chloride ions.

(ii) Describe how you could determine quantitatively the relative numbers of free chloride ions in A, B and C. **[10]**

(c) Suggest a shape for the cations present in A, B and C. **[1]**

(d) The chromium atom has the ground-state electron configuration $[Ar]3d^54s^1$.

(i) Write down the electron configuration of the Cr^{3+} ion.

(ii) Complete the diagram below to show the electronic structure of the cation $[Cr(H_2O)_6]^{3+}$, labelling clearly the electrons from the ligands.

	3d					4s	4p		
[Ar]									

(iii) What type of bond exists between the ligands and the chromium ion?

(iv) Name another neutral ligand which might replace water in this cation. **[5]**

32 The following question concerns the d-block elements, scandium to zinc.

(a) The highest oxidation state of the first six d-block elements are shown in the table below:

Element:	Sc	Ti	V	Cr	Mn	Fe
Highest oxidation state:	+3	+4	+5	+6	+7	+6

I/P

(i) Comment on the factors affecting the range of oxidation states shown by these elements. Suggest a value for the highest oxidation state for cobalt, explaining your choice.

(ii) State whether scandium should be considered a transition element, giving **two** reasons to support your answer. **[8]**

(b) Comment on and explain the stabilities of the Fe^{3+} and Mn^{3+} in aqueous solution in the light of the following standard electrode potentials.

$$Fe^{3+}(aq) + e^- \rightleftharpoons Fe^{2+}(aq) \qquad E^\ominus = +0.77 \text{ V}$$
$$Mn^{3+}(aq) + e^- \rightleftharpoons Mn^{2+}(aq) \qquad E^\ominus = +1.51 \text{ V}$$
$$O_2(g) + 4H^+(aq) + 4e^- \rightleftharpoons 2H_2O(l) \qquad E^\ominus = +1.23 \text{ V}. \qquad \textbf{[4]}$$

(c) Describe and explain what you would see when an aqueous solution of copper(II) sulphate was treated with aqueous ammonia solution until it was present in excess. Using examples from the reactions which occur, illustrate the meaning of the terms 'complex ion', 'ligand' and 'co-ordination number'. **[5]**

(d) A sample of steel (1.00 g) containing manganese was dissolved in nitric acid to give a solution containing manganese in oxidation state +2. All the manganese present was then oxidized to manganate(VII) by adding sodium bismuthate, $NaBiO_3$. After the destruction of any excess bismuthate ion, the resulting purple solution required 36.0 cm^3 of an iron(II) sulphate solution of concentration 0.10 mol dm^{-3} to reach an end-point, the iron(II) being oxidized to iron(III).

$$MnO_4^-(aq) + 8H^+(aq) + 5e^- \rightleftharpoons Mn^{2+}(aq) + 4H_2O(l)$$

(i) What is the oxidation state of bismuth in $NaBiO_3$?

(ii) Given that the bismuthate ion is reduced to Bi^{3+}, write an equation for the reaction in which it functions as an oxidizing agent. Hence write a balanced equation for the oxidation of manganese(II) to Mn(VII) by bismuthate in acidic solution.

(iii) Suggest why the oxidation was carried out in acidic solution.

(iv) Calculate the percentage by mass of manganese in the steel sample. **[12]**

33 A gaseous hydrocarbon, X, contains 88.89% C and 11.11% H. A simplified mass spectrum of X is shown on the next page.

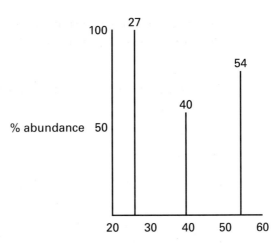

A

(a) (i) Calculate the empirical and molecular formulae of X.

(ii) Write down the possible isomers of X. **[5]**

In order to gain further information concerning the structure of X, it was reacted with hydrogen in the presence of a nickel catalyst: 20.0 cm³ of X reacted with exactly 40.0 cm³ hydrogen.

(b) (i) Write down the structure of X on the basis of this additional information.

(ii) Give the correct systematic name for X. **[3]**

(c) Identify the species responsible for the three peaks in the simplified mass spectrum of X. **[3]**

34* Methane and chlorine react in the presence of sunlight to form a number of chlorinated products, the first of which is chloromethane.

(a) (i) Use the following bond enthalpies to calculate the enthalpy change for the reaction:

$$CH_4(g) + Cl_2(g) \rightarrow CH_3Cl(g) + HCl(g)$$

(C—H 412, Cl—Cl 242, C—Cl 338, H—Cl 431, all in kJ/mol)

O

(ii) In the light of your answer, comment on the fact that the reaction does not occur in the dark. **[6]**

(b) Write down the propagation steps in the mechanism of the reaction. **[4]**

(c) Alternative propagation steps are as follows:

$$CH_4(g) + Cl\cdot(g) \rightarrow CH_3Cl(g) + H\cdot(g)$$
$$H\cdot(g) + Cl_2(g) \rightarrow Cl\cdot(g) + HCl(g)$$

(i) How might you tell from the products of the reaction that these propagation steps do not, in fact, occur?

(ii) Suggest why they do not occur. **[7]**

35 (a) (i) Draw and name the **three** structural isomers of the hydrocarbon of molecular formula C_5H_{12}.

(ii) Suggest why the boiling points of these isomers decrease as the hydrocarbon chain becomes more branched. **[10]**

(b) Two of the isomers shown in **(a)** become chiral when a chlorine atom is substituted for a hydrogen atom.

    (i) Draw **one** of these chiral isomers and its mirror image.

    (ii) What structural feature makes this isomer chiral? **[5]**

O/A

(c) The following questions concern a gaseous hydrocarbon, X, represented by the general formula C_xH_y. 10 cm³ of X were burned completely in 100 cm³ (an excess) of oxygen: 40 cm³ of carbon dioxide were formed and 40 cm³ of oxygen remained after the reaction. (All volumes were measured at room temperature and pressure and the volume of the water formed was negligible.)

    (i) Write a balanced equation for the complete combustion of C_xH_y in oxygen.

    (ii) What is the value of x?

    (iii) Calculate the value of y. **[8]**

(d) Identify the geometrical isomers of X and explain how such isomerism arises.

[3]

---

36 Methane undergoes homolytic free radical substitution with bromine in the presence of sunlight. Bromine contains about 50% each of ^{79}Br and ^{81}Br.

(a) **(i)** Explain the meaning of the terms 'homolytic' and 'free radical'.

    (ii) Why is sunlight necessary for the reaction to occur? **[4]**

(b) **(i)** For this reaction, write equations showing the initiation reaction and the propagation reactions.

    (ii) Suggest why the reaction is not generally used to prepare bromoalkanes.

[8]

O/A

(c) The mass spectrum of the reaction mixture was recorded and a part of it is reproduced below.

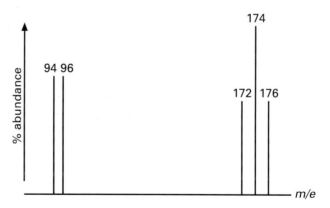

(i) Identify the species responsible for the peaks at m/e 94 and 96 and explain their relative intensities.

(ii) How many atoms of bromine are present in the bromoalkane which gives rise to the peaks between m/e 172 and 176? Give your reasoning.

(iii) There is a peak in the mass spectrum at m/e 30. Suggest the species responsible for it and explain how its occurrence amongst the products supports the proposed reaction mechanism. **[10]**

37 Use the information below to answer the questions which follow.

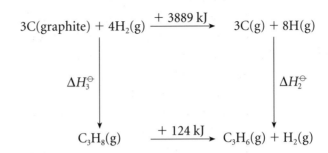

$$3C(\text{graphite}) + 4H_2(g) \xrightarrow{+\ 3889\ kJ} 3C(g) + 8H(g)$$

$$\Delta H_3^{\ominus} \qquad\qquad \Delta H_2^{\ominus}$$

$$C_3H_8(g) \xrightarrow{+\ 124\ kJ} C_3H_6(g) + H_2(g)$$

P

Standard enthalpy change of atomization of C(graphite) $= +715$ kJ mol^{-1}
Average bond enthalpies: C—C $= +346$ kJ mol^{-1}, C—H $= +413$ kJ mol^{-1}

(a) On what law is the above cycle based? [1]

(b) What name is given to the enthalpy change $\Delta H_3^{\ominus}$? [2]

(c) Calculate a value for the standard enthalpy change of atomization of hydrogen. [3]

(d) Calculate a value for the enthalpy change of atomization of propane, $C_3H_8(g)$. [2]

(e) (i) Using the values from **(c)** and **(d)** and any other data you require, draw an energy cycle to calculate the standard enthalpy of atomization of propene, $C_3H_6(g)$. Then calculate a value for this energy change.

(ii) Use your answer to **(e)(i)** to calculate a value for the mean bond enthalpy of the C=C bond. [8]

38 (a) (i) Give the reagents and conditions required to prepare the Grignard reagent from bromobenzene, C_6H_5Br, in the laboratory.

(ii) Write an equation for the formation of this Grignard reagent.

(iii) What safety precaution is particularly important when preparing Grignard reagents? Why? [5]

(b) (i) Write down the structural formulae of the organic substances with which this Grignard reagent must be reacted to give A, B, C and D below.

O

A
$$\begin{array}{c} C_6H_5 \\ | \\ H-C-OH \\ | \\ H \end{array}$$

B
$$\begin{array}{c} C_6H_5 \\ | \\ H-C-OH \\ | \\ CH_3 \end{array}$$

C
$$\begin{array}{c} C_6H_5 \\ | \\ CH_3-C-OH \\ | \\ CH_3 \end{array}$$

D $C_6H_5-CH_2-CH_2-OH$

(ii) Describe a chemical test which would give a positive result only with B.

(iii) Explain how A and B might be distinguished without the need for a chemical test. **[12]**

39 (a) Methylbenzene can be nitrated with a mixture of concentrated nitric and sulphuric acids. The reaction is classified as electrophilic substitution.

(i) Explain what is meant by the term 'electrophile'.

(ii) Give the formula of the electrophile in the nitration of methylbenzene.

(iii) Suggest, by means of equations, how this electrophile might be formed in the reaction between concentrated nitric and sulphuric acids.

(iv) Show the mechanism of the reaction between methylbenzene and this electrophile to give 2-nitromethylbenzene.. **[7]**

(b) In the presence of anhydrous aluminium chloride, the reaction between methylbenzene and chlorine is an electrophilic substitution.

(i) Write an equation showing the reaction between aluminium chloride and chlorine and underline the electrophile formed in the reaction.

(ii) Write down the name and formula of the organic substance with which methylbenzene would react in the presence of aluminium chloride to give a product with a methylketone side-chain, $CH_3—CO—$. **[4]**

(c) The side chain in methylbenzene can be oxidised by acidified potassium manganate(VII) which is reduced to aqueous Mn^{2+} ions.

(i) Name the organic product formed when the side chain is oxidised.

(ii) Write down the formula of potassium manganate(VII).

(iii) Write a balanced half-equation for the reduction of the manganate(VII) ion to Mn^{2+}.

(iv) Use your equation to define the term 'reduction'. **[6]**

40 The reaction scheme below represents the conversion of benzene to phenylamine.

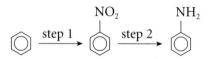

(a) (i) Give the reagents and conditions required for step 1 and step 2.

(ii) Write down the mechanism for the reaction occurring in step 1. **[9]**

(b) Phenylamine is insoluble in water but dissolves readily in dilute hydrochloric acid. Treatment of the resulting solution below 10 °C with aqueous sodium nitrite gives a solution containing a compound X. A precipitate forms when the solution containing X is made alkaline and then phenol is added.

(i) Give the formula of the species formed when phenylamine dissolves in dilute hydrochloric acid.

(ii) Suggest why phenylamine is more soluble in dilute hydrochloric acid than it is in water.

(iii) Write down the formula of compound X.

(iv) If aqueous solutions of X are warmed, X is hydrolysed, a gas being among the products. Write a balanced equation for the hydrolysis of X.

(v) What **type** of reaction takes place between X and phenol in alkaline solution?

(vi) Write down the structural formula for the precipitate formed when X reacts with phenol in alkaline solution and suggest what colour it might be.

[15]

41 **(a)** State and explain how each of the following would affect the rate of a chemical reaction taking place in solution:

(i) an increase in temperature

(ii) the addition of a catalyst

(iii) an increase in the concentration of one of the reactants.

[13]

(b) The table below shows some experimental data for the reaction

$$A + B \rightarrow C.$$

All concentrations are in mol dm^{-3}.

P

[A]	[B]	initial rate/mol dm^{-3} s^{-1}	[A]	[B]	initial rate/mol dm^{-3} s^{-1}
1.0	1.0	1.2×10^{-2}	1.0	1.0	1.2×10^{-2}
2.0	1.0	2.4×10^{-2}	1.0	2.0	4.8×10^{-2}
4.0	1.0	4.9×10^{-2}	1.0	4.0	1.9×10^{-1}

State, showing clearly how you arrived at your answer, the order with respect to:

(i) reactant A

(ii) reactant B.

(iii) Write down the rate law expression for the reaction.

(iv) What is the overall order of the reaction?

(v) Calculate the value of the rate constant using the set of data for [A] = 1.0 mol dm^{-3} and [B] = 2.0 mol dm^{-3} in the table. Give the units of the rate constant.

[11]

42 **(a) (i)** Draw the structural formula of a chiral molecule having the molecular formula C_4H_9Cl and mark the chiral carbon atom with an asterisk (*).

(ii) What property does this molecule possess?

[4]

O

(b) Ethane and chlorine react when a mixture of the two is exposed to a bright light source.

(i) What **type** of reaction takes place under these conditions?

(ii) Give equations for: the initiation reaction, the propagation reactions, a termination reaction.

[9]

43 Benzene can be converted to methylbenzene in the presence of an aluminium chloride catalyst. The reaction is classified as electrophilic substitution.

(a) **(i)** Give the name and formula of the organic reagent required in the reaction.

(ii) Give the formula of the electrophile in this reaction.

(iii) Give an equation to show how this electrophile might be formed in the reaction between the organic reagent in **(i)** and aluminium chloride.

O

(iv) Show the mechanism of the reaction between benzene and this electrophile to give methylbenzene, showing how the aluminium chloride behaves as a catalyst. **[9]**

(b) For benzene and ethene give **one** example of a chemical difference and **one** example of a chemical similarity. **[2]**

(c) A compound of molecular formula C_8H_{10} contains a benzene ring. Write down the structural formulae of the isomers of the compound. **[3]**

44 **(a)** Explain what is meant by each of the following terms:

(i) order of a chemical reaction

(ii) rate constant. **[4]**

(b) The following initial rates of reaction for the hydrolysis of methyl ethanoate in aqueous hydrochloric acid were measured at 30 °C.

	initial concentrations/mol dm^{-3}	
initial rate/mol dm^{-3} min^{-1}	methyl ethanoate	hydrochloric acid
0.001	1.0	0.5
0.002	1.0	1.0
0.001	0.5	1.0

P

(i) Give the order of the reaction with respect to:
   hydrogen ions
   methyl ethanoate.

(ii) Give the overall order of reaction

(iii) Write the rate equation for this reaction.

(iv) Calculate the value of the rate constant, stating its units. **[7]**

(c) **(i)** What is meant by the term 'activation energy'?

(ii) Copy the axes on the right and sketch the distribution of molecular energies for a reaction at a temperature of T °C and (T + 500) °C, labelling each line.

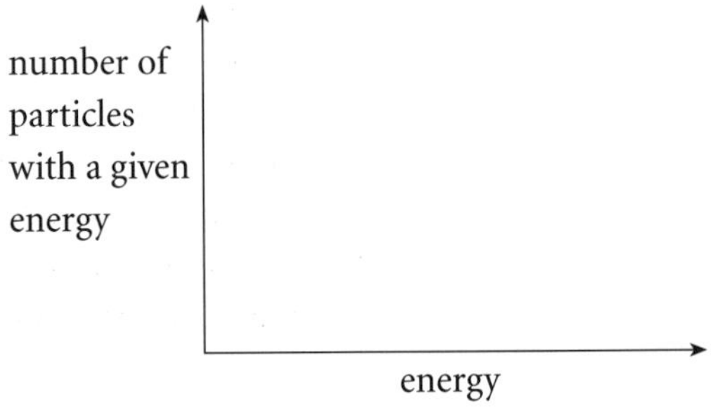

number of particles with a given energy

energy

(iii) Use your sketch to explain the effect of increasing the temperature on the rate of a chemical reaction and the effect of a catalyst on the rate of a chemical reaction. **[9]**

45* Ethene reacts readily with bromine in the dark, but the reaction with ethane occurs only in sunlight.

(a) (i) Show the mechanism of the reaction between ethene and bromine.

(ii) How can the enhanced reactivity of ethene be explained?

(iii) What **type** of reaction occurs between ethene and bromine? **[9]**

(b) When ethene reacts with bromine in an aqueous solution of sodium chloride a compound of molecular formula C_2H_4BrCl is among the products.

(i) Explain how this product is formed.

(ii) Explain why a product of molecular formula $C_2H_4Cl_2$ is not formed in this reaction. **[6]**

(c) (i) Draw the structures of **two** different products which might result when propene reacts with hydrogen bromide.

(ii) Explain, in terms of the mechanism of the reaction, why one of these two products is formed in greater yield than the other. **[11]**

46* The chiral halogenoalkane 2-bromobutane undergoes nucleophilic substitution on reaction with aqueous sodium hydroxide at 30 °C to give an optically active reaction mixture, but if the reaction is repeated at 80 °C the reaction mixture shows hardly any optical activity. The reaction between 2-bromo-2-phenyl butane (C_6H_5—$CBr(CH_3)$—CH_2CH_3) yields a racemic mixture whatever the temperature. If ethanolic solutions of sodium hydroxide are used, elimination reactions occur with both halogenoalkanes to give alkenes.

(a) (i) Draw the structural formula of 2-bromobutane.

(ii) What is meant by the term 'chiral'?

(iii) What feature of the structure of 2-bromobutane makes it chiral? **[4]**

(b) (i) Write the mechanism for the reaction between 2-bromobutane and aqueous sodium hydroxide at 30 °C.

(ii) Explain why the resulting reaction mixture shows optical activity.

(iii) Suggest why the reaction mixture obtained when 2-bromobutane and aqueous sodium hydroxide react at 80 °C shows hardly any optical activity. **[9]**

(c) (i) Give **two** reasons why the reaction between 2-bromo-2-phenylbutane and sodium hydroxide proceeds via an S_N1 mechanism rather than S_N2.

(ii) Explain, in terms of the mechanism, why this reaction results in a racemic mixture of products. **[7]**

(d) Give the structural formula and names of the products which result from the elimination reaction involving 2-bromobutane and 2-bromo-2-phenylbutane. **[10]**

47 **(a)** 72 cm^3 of ammonia gas were reacted with an equal volume of hydrogen chloride gas under normal laboratory conditions. What mass of ammonium chloride could be formed? The equation for the reaction is:

$$NH_3(g) + HCl(g) \rightarrow NH_4Cl(s)$$ [2]

I/O

(b) What mass of ethanal, CH_3CHO, could theoretically be obtained from one cubic decimetre of a 5% solution of ethanol, C_2H_5OH? (A 5% solution contains 5 g of solute in 100 cm^3 of solution.) A simplified equation for the reaction is:

$$C_2H_5OH + [O] \rightarrow CH_3CHO + H_2O$$ [2]

(c) Sulphur dioxide can be removed from the effluent gases produced by a power station by reacting it with limestone (calcium carbonate).

$$2CaCO_3(s) + 2SO_2(g) + O_2(g) \rightarrow 2CaSO_4(s) + 2CO_2(g)$$

It has been estimated that 3 million tonnes of sulphur dioxide are produced per year by power stations in the United Kingdom. How much limestone would theoretically be needed to remove all of this sulphur dioxide? [2]

(d) Barium ferrate(VI), $BaFeO_4$, can be produced by the procedure summarized in the equations below:

$$2Fe(NO_3)_3 + 3NaOCl + 10NaOH \rightarrow 2Na_2FeO_4 + 6NaNO_3 + 3NaCl + 5H_2O$$

followed by:

$$Na_2FeO_4(aq) + Ba(NO_3)_2(aq) \rightarrow BaFeO_4(s) + 2NaNO_3(aq)$$

Starting with 1.00 g of iron(III) nitrate-9-water, $Fe(NO_3)_3 . 9H_2O$, and an excess of any other reagent required, what is the maximum mass of barium(VI) ferrate which could be obtained? [3]

(e) Potassium trioxalatocobaltate(III), $K_3[Co(C_2O_4)_3]$, can be made by a series of steps summarized in the equation below:

$$2CoCO_3 + 3H_2C_2O_4 + 3K_2C_2O_4 + PbO_2 + 2CH_3CO_2H \rightarrow$$
$$2K_3[Co(C_2O_4)_3] + (CH_3CO_2)_2Pb + 4H_2O + 2CO_2$$

To make 10.0 g of potassium trioxalatocobaltate(III), what are the minimum masses of cobalt(II) carbonate, ethanedioic acid-2-water, $H_2C_2O_4.2H_2O$, potassium ethanedioate-1-hydrate, $K_2C_2O_4.H_2O$ and lead(IV) oxide which would be required? [6]

48 **(a)** Triphenylmethanol can be made from benzoic acid, $C_6H_5CO_2H$, by the sequence of reactions shown below (not all products or reactants are shown):

$3CH_3OH$

$+ \qquad \rightarrow 3C_6H_5CO_2CH_3 \rightarrow (C_6H_5)_3COMgBr \rightarrow (C_6H_5)_3COH$

$3C_6H_5CO_2H$

O/I

In carrying out the series of reactions, 10.0 g of benzoic acid was used initially. Reacting this with an excess of methanol, CH_3OH, this yielded 8.0 g of methyl benzoate. Then using this quantity of the first product, an impure magnesium salt was produced. Hydrolysis followed by purification of the product resulted in a final yield of 4.0 g. What was the percentage yield in:

(i) the first step [2]

(ii) the second and third steps, taken together [2]

(iii) the overall series of reactions? [1]

(b) Trisodiumhexanitrocobaltate(III) crystals have the formula $Na_3[Co(NO_2)_6]$. They can be prepared from cobalt(II) nitrate-6-water, $Co(NO_3)_2 \cdot 6H_2O$, and an excess of sodium nitrite in the presence of dilute ethanoic acid. No other cobalt-containing product is formed.

A student started with 3.0 g of cobalt(II) nitrate-6-water and finished with 2.5 g of trisodiumhexanitrocobaltate(III). What was the percentage yield? [3]

49 In each part of this question, you must begin by writing a balanced equation for the reaction. Give your answers to parts (a)–(f) inclusive, correct to two significant figures and your answers to parts (g) and (h) correct to three significant figures.

(a) 22.5 cm³ of a solution of 0.1 M hydrochloric acid react with 25.0 cm³ of a solution of sodium hydroxide. What is the concentration of the sodium hydroxide solution? [3]

(b) 10.0 cm³ of a solution of 0.05 M sodium hydroxide solution react with 8.8 cm³ of a solution of hydrochloric acid. What is the concentration of the acid? [2]

(c) 32.2 cm³ of a 0.20 M solution of sulphuric acid react with 25.0 cm³ of a solution of potassium carbonate. What is the concentration of the carbonate solution? [3]

I/A

(d) 16.0 cm³ of a solution of barium hydroxide react with 20.0 cm³ of a solution of 0.25 M solution of hydrochloric acid. What is the mass of barium hydroxide dissolved in 1 dm³ of the solution? [4]

(e) 7.6 cm³ of the dibasic acid, ethanedioic acid, $H_2C_2O_4$, react exactly with 10.0 cm³ of 0.08 M sodium hydroxide solution. Calculate the concentration of the ethanedioic acid. [3]

(f) 10.0 cm³ of arsenic(V) acid, H_3AsO_4, react with 14.2 cm³ of a 0.10 M solution of sodium hydroxide. Assuming that all the hydrogen in the acid reacts, calculate the concentration of the acid in (i) mol dm⁻³ and (ii) g dm⁻³. [4]

(g) A solution of hydrochloric acid of concentration 0.15 M is to be titrated against 25.0 cm³ of a solution of 0.125 M sodium hydroxide solution. What volume of acid is likely to be needed? [2]

(h) A solution of 0.35 M ammonia solution is to be titrated against 10.0 cm³ of a 0.25 M solution of sulphuric acid. What volume of ammonia solution is likely to be needed? [2]

50 Borax, sodium borate $Na_2B_4O_7.10H_2O$, is, according to some chemists, the most reliable substance with which to standardise a solution of an acid. It reacts with acid forming boric acid, which is so weak an acid that virtually none of it is dissociated into ions and therefore does not interfere with the titration if the correct indicator is chosen. This reaction can be summarized by the following equation:

I/A

$$B_4O_7^{2-} + 2H^+ + 5H_2O \rightarrow 4H_3BO_3$$

1.15 g of borax was dissolved in warm distilled water and made up to 100 cm^3. 10.0 cm^3 portions of the borax solution were titrated with the sulphuric acid whose concentration has to be determined accurately. 20.0 cm^3 of acid was required.

(a) What is the concentration of the borax solution in mol dm^{-3}? **[2]**

(b) How many moles of borax are present in 10.0 cm^3 of solution? **[1]**

(c) How many moles of hydrogen ions must therefore be present in the acid used? **[1]**

(d) What is the concentration of the sulphuric acid? **[1]**

51 0.55 g of an impure sample of ammonium chloride was warmed with 40.0 cm^3 of 1.00 M sodium hydroxide solution.

$$NH_4Cl(s) + 2NaOH(aq) \rightarrow 2NH_3(g) + 2NaCl(aq) + 2H_2O(l)$$

I/A When all the ammonia had been given off, the excess sodium hydroxide was titrated with hydrochloric acid of concentration 0.75 mol dm^{-3}. It was found that 26.7 cm^3 was required for neutralization.

$$NaOH(aq) + HCl(aq) \rightarrow NaCl(aq) + H_2O(l)$$

Calculate the percentage of ammonium chloride present in the sample. **[6]**

52 A geologist wanted to determine the percentage purity of a sample of limestone (calcium carbonate) by reacting it with sulphamic acid, H_3SO_3N. However, he did not know how many of the hydrogen atoms had acidic properties and so decided to investigate this by reacting sulphamic acid with sodium carbonate solution.

I/A He made up a solution of sulphamic acid containing 1.50 g in 250 cm^3 of solution and titrated a 20.0 cm^3 sample with 0.025 M sodium carbonate using a suitable indicator. 24.8 cm^3 of sodium carbonate solution was required for complete reaction.

Sodium carbonate and hydrogen ions react as indicated in the equation below:

$$CO_3^{2-} + 2H^+ \rightarrow H_2O + CO_2$$

Calculate the number of moles of replaceable hydrogen atoms in one mole of sulphamic acid. **[6]**

53 The substance known as 'microcosmic salt' is hydrated ammonium sodium hydrogen phosphate and has the formula $NH_4NaHPO_4.xH_2O$.

I/A In order to find the value of x, a student weighed out 10.0 g of microcosmic salt and dissolved it in distilled water in a 100 cm^3 volumetric flask before making it up to the mark. She then took 20.0 cm^3 of this solution and added the same volume of 1.0 M sodium hydroxide. The mixture was boiled gently until no more ammonia vapour could be detected. After cooling, the mixture was titrated with 0.5 M hydrochloric acid using phenolphthalein as indicator. The process was repeated and the average of two concordant titres was obtained – 14.2 cm^3.

The essential reaction is:

$$NH_4^+ + OH^- \rightarrow NH_3 + H_2O$$

Calculate the value of x. [9]

54 Aspirin has the systematic name 2-ethanoyloxybenzoic acid and the structural formula $CH_3CO_2C_6H_4CO_2H$. Commercial aspirin tablets contain inactive ingredients as well as 2-ethanoyloxybenzoic acid. To find how much of the latter is present, the tablets can be simmered for 10 minutes with 1.0 M sodium hydroxide solution which hydrolyses the aspirin in the tablets forming sodium 2-hydroxybenzoate. The amount of sodium hydroxide solution remaining is then found by titration, from which information the amount of 2-ethanoyloxybenzoic acid originally present can be determined.

$$CH_3CO_2C_6H_4CO_2H + 2NaOH \rightarrow HOC_6H_4CO_2Na + CH_3CO_2Na + H_2O$$

O/I/A In a particular experiment 10 tablets were added to 25.0 cm^3 of 1.0 M sodium hydroxide solution. After hydrolysis, the solution was made up to the mark in a 250 cm^3 standard flask. 25.0 cm^3 pipette volumes of this solution were titrated with 0.05 M sulphuric acid, using a suitable indicator. It was found that 8.0 cm^3 were required for neutralization of the excess alkali present. Calculate the mass of 2-ethanoyloxybenzoic acid present in each tablet using the following steps.

(a) How many moles of sodium hydroxide were used initially? [1]

(b) Write the equation for the reaction between sodium hydroxide and dilute sulphuric acid. [1]

(c) How many moles of sodium hydroxide were left in the standard flask at the end of the hydrolysis? [1]

(d) How many moles of 2-ethanoyloxybenzoic acid were therefore hydrolysed? [1]

(e) What mass of ethanoyl 2-ethanoyloxybenzoic acid was therefore present in **each** tablet? [1]

(f) Name a suitable indicator for the titration. [1]

55

O/P

Propene has a melting point of $-185\,°C$ and a boiling point of $-47\,°C$ while poly(propene) has a melting point of about $175\,°C$ and is a widely used plastic with many uses such as rope, bottle crates and chairs. Discuss this in relation to the type and arrangement of their molecules. [4]

56 **(a)** Find the standard enthalpy change of reaction for:

$$2KHCO_3(s) \rightarrow K_2CO_3(s) + CO_2(g) + H_2O(l)$$ [2]

You will need the following data:

$$\Delta H_f^{\ominus}[KHCO_3(s)] = -963.2\,kJ\,mol^{-1}$$
$$\Delta H_f^{\ominus}[K_2CO_3(s)] = -1151.0\,kJ\,mol^{-1}.$$

I/P

(b) Find the standard enthalpy change of reaction for the corresponding reaction involving sodium hydrogencarbonate. [1]
The standard enthalpy changes of formation needed are:

$$\Delta H_f^{\ominus}[NaHCO_3(s)] = -950.8\,kJ\,mol^{-1}$$
$$\Delta H_f^{\ominus}[Na_2CO_3(s)] = -1130.7\,kJ\,mol^{-1}.$$

(c) Write the ionic equations for each of the two reactions. In the light of this, comment on the values you have obtained in **(a)** and **(b)**. [2]

57 Find the standard enthalpy change of reaction for the esterification:

$$CH_3CO_2H(l) + C_2H_5OH(l) \rightarrow CH_3CO_2C_2H_5(l) + H_2O(l)$$ [2]

O/P

The standard enthalpy changes of formation required are:

$$\Delta H_f^{\ominus}[CH_3CO_2H(l)] = -484.5\,kJ\,mol^{-1}$$
$$\Delta H_f^{\ominus}[C_2H_5OH(l)] = -277.1\,kJ\,mol^{-1}$$
$$\Delta H_f^{\ominus}[C_2H_5CO_2C_2H_5(l)] = -479.3\,kJ\,mol^{-1}.$$

Explain what is unusual about the value of the answer obtained. [1]

58 $10\,cm^3$ of a saturated solution of potassium bromide was added to $30\,cm^3$ of a solution of copper(II) sulphate of concentration $0.20\,mol\,dm^{-3}$. It turned from blue to green. The reaction involved was:

$$CuSO_4(aq) + 4KBr(aq) \rightleftharpoons K_2[CuBr_4](aq) + K_2SO_4(aq)$$

I/P

The cation actually present in the copper sulphate solution is $[Cu(H_2O)_6]^{2+}(aq)$, a complex ion which gives it its characteristic blue colour. Copper can form many such complex ions e.g. $[CuCl_4]^{2-}(aq)$ which is yellow, formed as in the following reaction:

$$[Cu(H_2O)_6]^{2+}(aq) + 4Cl^-(aq) \rightleftharpoons [CuCl_4]^{2-}(aq) + 6H_2O(l)$$

(a) Explain the following observations:
  (i) Solid sodium sulphate, Na_2SO_4, was added to $10\,cm^3$ of the green solution. It turned blue. [2]

(ii) The remaining solution was then left to stand for some time until it became blue-green and was then divided into three portions, (A), (B) and (C).

Portion (A) was warmed. It turned blue.

Portion (B) was placed in an ice-bath. It turned green.

Portion (C) was left at room temperature. It stayed blue-green. **[3]**

(b) From the results described in **(a)(ii)**, what can you deduce about the nature of the energy change for the forward reaction shown in the equation above? **[1]**

(c) What colour changes (if any) would you expect to see if **(i)** a large volume of sodium chloride solution were to be added to portion (A) after it had been warmed and **(ii)** the solution were subsequently to be diluted? Explain your answers. **[4]**

59 An aqueous solution of cobalt(II) chloride is pink; a result of its containing the $[Co(H_2O)_6]^{2+}(aq)$ ion. On the addition of concentrated hydrochloric acid, the solution turns blue as a result of the reaction:

$$[Co(H_2O)_6]^{2+}(aq) + 4Cl^-(aq) \rightleftharpoons [CoCl_4]^{2-}(aq) + 6H_2O(l)$$

I/P

Use is made of this product in colouring the drying agent, silica gel, which is used, for example, in desiccators. When the drying agent is active, it is blue. When it is saturated with water the colour changes to pink. It is then converted back to blue on heating in an oven at 120 °C. By discussing the effect of concentration changes on the essential reaction involved, explain these observations. **[4]**

60 Sodium iodide solution of concentration 0.1 mol dm^{-3} was added to copper(II) sulphate solution of the same concentration, in a separating funnel, until a yellowish-brown precipitate just formed. The mixture was then shaken with cyclohexane, a solvent for iodine molecules. Cyclohexane is immiscible with water but a good solvent for iodine. The upper organic layer was purple in colour while the lower aqueous layer turned golden brown.

The following experiments were then carried out:

I/P

(a) A very small amount of concentrated ammonia solution was added to a very small portion of the aqueous layer. It decolorised the latter.

(b) Excess dilute ammonia solution was then added to the separating funnel and the mixture was shaken vigorously. The organic layer was decolourised and the aqueous layer also lost its colour.

(c) Sulphuric acid, of concentration 6 mol dm^{-3}, was then added to the separating funnel until in excess and the mixture was shaken vigorously. The aqueous layer became golden brown and the organic layer purple.

Try to explain these observations. **[4]**

61 Write down the correct answers corresponding to **(a)–(o)** in the table below.

symbol for isotope	atomic number	number of protons	number of neutrons	mass number	
(a)	17	(b)	20	(c)	[3]
$^{58}_{28}Ni$	(d)	28	(e)	(f)	[3]
(g)	(h)	29	(i)	65	[3]
(j)	(k)	34	40	(l)	[3]
(m)	35	(n)	(o)	81	[3]

62 **(a)** Describe in outline how a mass spectrometer functions. [7]

  (b) How many peaks would you expect to see in the parent ion region of the mass spectrum of the following compounds? The stable isotopes of the elements involved are shown. [8]

molecule	stable isotopes of the elements in the compound
$BeCl_2$	$^{9}_{4}Be$ (100%), $^{35}_{17}Cl$ (75%), $^{37}_{17}Cl$ (25%)
PCl_3	$^{31}_{15}P$ (100%), $^{35}_{17}Cl$ (75%), $^{37}_{17}Cl$ (25%)
HCN	$^{1}_{1}H$ (very nearly 100%), $^{12}_{6}C$ (98.9%), $^{13}_{6}C$ (1.1%), $^{14}_{7}N$ (99.6%), $^{15}_{7}N$ (0.4%)
ClO_2	$^{16}_{8}O$ (99.8%), $^{18}_{8}O$ (0.2%), $^{35}_{17}Cl$ (75%), $^{37}_{17}Cl$ (25%)

63 Five elements, denoted by the letters A to E, have the electronic structures shown:

A $1s^2\,2s^2\,2p^6\,3s^2\,3p^1$
B $1s^2\,2s^2\,2p^6\,3s^2\,3p^6\,4s^2$
C $1s^2\,2s^2\,2p^6\,3s^2\,3p^6$
D $1s^2\,2s^2\,2p^6\,3s^2\,3p^5$

E $1s^2\,2s^2\,2p^6\,3s^2\,3p^6\,3d^5\,4s^1$

In each case, choose from A to E an element which fits the description given.
(a) It has an oxidation state of $+2$ in all its compounds.
(b) It is chemically unreactive.
(c) It is a powerful oxidizing agent.
(d) It forms two oxides, both of which are coloured.
(e) It forms a $+3$ ion and has a chloride which sublimes easily. [5]

64 **(a)** Explain the difference in boiling points of each of the following pairs of
substances in terms of the bonding in each member of the pair.
  (i) NaCl 1465 °C and HCl −85 °C [3]
  (ii) H_2O 100 °C and CH_4 −162 °C [3]
  (iii) I_2 184 °C and F_2 −188 °C [3]

P

(b) Explain the differences in boiling points between the following pairs of
substances in terms of their bonding and structure.
  (i) pentane 36 °C and 2,2-dimethylpropane 10 °C [3]
  (ii) butan-1-ol 117 °C and ethoxyethane, $C_2H_5-O-C_2H_5$, 35 °C [3]
  (iii) potassium 776 °C and copper 2567 °C [3]
  (iv) methane −162 °C and ammonia −33 °C [3]

65 **(a)** Give **one** example of each of the following types of solids:
simple molecular, giant covalent, ionic. [3]

P

(b) Explain, in terms of the structure:
  (i) the hardness of most giant covalent substances [3]
  (ii) the inability of simple molecular substances to conduct electricity [3]
  (iii) the tendency of ionic substances to shatter when given a sharp blow. [3]

66 Explain the following in terms of the intermolecular forces present:

P

(i) Xenon boils at −108 °C, while argon boils at −186 °C. [2]
(ii) Hydrogen fluoride is a liquid at 15 °C, while hydrogen chloride is gaseous at
this temperature. [3]

67 Write the names of:
(a) CH_3CH_2CHO [1]

(b) **(i)** $CH_3CCH_2CH_3$
         $\overset{\|}{O}$ [1]

O

(ii) Explain why there is no need for any number to be included in the
name. [1]

(c) $CH_3CCH_2CH_2OH$ [1]   **(d)** $CH_3CCH_2CH_2Br$ [1]
      $\overset{\|}{O}$ $\overset{\|}{O}$

                                                         CH_3
                                                          $|$
(e) $CH_3C\ CHCH_2OH$ [1]   **(f)** CH_3CCHO [1]
      $\overset{\|}{O}\ \overset{|}{Cl}$ $\overset{|}{OH}$

(g) **(i)** Write the structural formula of the ketone which has a molecular
formula of C_3H_6O. [1]
(ii) Write the structural formula of the straight chain aldehyde which has a
molecular formula of C_4H_8O. [1]
(iii) Write the formula of the other isomer of the aldehyde C_4H_8O. [1]

139

68 **(a)** Lactic acid is produced by anaerobic respiration in human cells. It is the cause of aching muscles. The formula for lactic acid is:

$$CH_3CHCOOH$$
$$|$$
$$OH$$

O   What is the systematic name for lactic acid?   **[1]**

    (b) Pyruvic acid is a ketone obtained from lactic acid. It may be named keto-propanoic acid. Write the structural formula for pyruvic acid.   **[1]**

    (c) **(i)** Oxalic acid, or ethanedioc acid, is a dicarboxylic acid. It contains only two carbon atoms. Write the structural formula for oxalic acid.   **[1]**

        (ii) The next dicarboxylic acids in the homologous series are malonic acid and succinic acid. They contain three and four carbon atoms respectively. Write the structural formulae for these two acids.   **[2]**

69 Prepare a table showing the empirical, molecular and structural formulae for the classes of compounds listed below. In each case set up the table to include two carbon and three carbon compounds in the empirical and molecular formula columns and show any isomers of the three carbon compounds in the structural formula column. For information:

O   the empirical formula for ethene is CH_2 and for propene is CH_2

    the molecular formula for ethene is C_2H_4 and for propene is C_3H_6

    the structural formula for propene is $CH_3CH{=}CH_2$.

    (a) alkanes   **(b)** alcohols   **(c)** amines

    (d) chloroalkanes   **(e)** ketones   **(f)** aldehydes

    (g) carboxylic acids   **(h)** amides   **[8 × [3]]**

70* There are three ways to detect a gaseous alkene such as ethene. They are:

    A   Bubble the gas through bromine dissolved in an organic solvent such as dichloroethane.

    B   Bubble the gas through bromine water.

O   C   Bubble the gas through acidified potassium manganate(VII) solution.

    (a) In each case describe what is seen, name the organic product which may be formed from ethene and write its structural formula.   **[9]**

    (b) State the names of the possible products formed if the gas is but-1-ene rather than ethene.   **[4]**

71* **(a)** Write the formula of the ester formed by the reaction of ethanol with methanoic acid.   **[1]**

O   **(b)** Use curly arrows to show the mechanism for the formation of the ester. Discuss the reasons for the mechanism operating in the manner you show.   **[4]**

72* **(a) (i)** Use a 3D convention to draw the two optical isomers of lactic acid (2-hydroxypropanoic acid). [1]

(ii) Describe how a solution of each isomer affects polarized light. [1]

O

(b) (i) Pyruvic acid is obtained by oxidizing lactic acid. Write the structural formula for pyruvic acid. [1]

(ii) Explain why lactic acid obtained by reducing the ketone function in pyruvic acid is racemic. [1]

73 Three types of organic chloro-compound are shown by:

CH_3CH_2Cl CH_3COCl Cl

O

(a) Give the name of each compound. [3]

(b) (i) State if each reacts with water, and write the formula for any product. [3]

(ii) State the conditions under which each reacts with NaOH and what is formed. [7]

(c) Explain why each compound behaves differently toward nucleophiles. [12]

74 A systematic name for aspirin is 2-ethanoyloxybenzoic acid. Its formula is $CH_3CO_2C_6H_4CO_2H$ and it has a K_a value of 3.0×10^{-4} mol dm^{-3}. As well as being used as an analgesic, aspirin is taken regularly by some people in order to reduce the likelihood of suffering from heart disease. A particular patient is recommended to take, on a daily basis, a tablet containing 0.15 g of aspirin and this is ingested after dissolving in 25 cm^3 of water at room temperature.

(a) Calculate the following assuming that all the aspirin has dissolved in water:

(i) the hydrogen concentration $[H^+(aq)]$ [4]

(ii) the pH. [1]

P

(b) The mouth has a pH of around 7 and the stomach a pH of around 2. Explain what happens to the extent of dissociation of aspirin molecules into ions as the substance passes from the mouth to the stomach. [2]

(c) The changes in the aspirin molecules taking place in the stomach allow it to reach the stomach wall where it encounters a higher pH, close to neutrality. Account for the possibility of a gastric bleeding side-effect caused by the attack of $H^+(aq)$ ions upon the stomach side wall. [2]

(d) The more soluble a medicine is, the faster is the rate of passage from the digestive system into the bloodstream and so the quicker its action. Aspirin is not very soluble in water. Suggest how it might be readily converted into a more soluble form. [2]

75 Citric acid (2-hydroxypropane-1,2,3-tricarboxylic acid) has the formula:

$$HO_2C - \overset{\displaystyle H}{\underset{\displaystyle H}{C}} - \overset{\displaystyle OH}{\underset{\displaystyle CO_2H}{C}} - \overset{\displaystyle H}{\underset{\displaystyle H}{C}} - CO_2H$$

P

(a) Explain why **three** values of K_a are given in data books for citric acid: 7.10×10^{-3}, 1.68×10^{-5} and 6.40×10^{-6} (all in mol dm^{-3}). **[3]**

(b) What are the equivalent pK_a values for citric acid? **[2]**

(c) Write a balanced equation for the complete neutralization of citric acid by sodium hydroxide solution. **[2]**

(d) Citric acid is present in lemon juice. Refrigerated lemon juice tastes less sour than unrefrigerated lemon juice of the same concentration. Explain this. In addition, show that this observation is consistent with the ionization of citric acid being an endothermic process. **[3]**

76 When a mixture of calcium oxide and silicon dioxide is strongly heated a crystalline product with the following composition by mass is obtained: Ca 34.5%, Si 24.1%, O 41.4%.

I

(a) (i) Calculate the formula of the crystalline product. **[2]**

(ii) Suggest formulae for the ions which it contains. **[2]**

(b) The solid reacts with neither acids nor alkalis. Suggest why. **[3]**

77 Suggest explanations for the following.

(a) Of the first-row transition elements, manganese reacts most readily with acids. **[3]**

I/P

(b) The maximum oxidation state for the elements Sc to Mn is equal to the total number of 3d and 4s electrons. If this trend were followed by iron, its maximum oxidation state would be +8, but instead it is +6. **[3]**

(c) Despite their identical outer electronic configurations ($4s^1$) potassium and copper differ greatly in their reactivity with water. **[5]**

78 The treatment of a mixture of potassium chloride and copper(I) chloride with fluorine results in the formation of a substance having the formula K_3CuF_6.

(a) (i) Give the formula of the complex ion in this substance. **[1]**

(ii) What shape would the complex ion have? Explain your answer. **[2]**

I/P

(b) (i) What is the oxidation state of copper in the complex ion? **[1]**

(ii) Write down the electronic configuration of copper in this oxidation state. **[1]**

(iii) Suggest **two** properties which you would expect this oxidation state of copper to have. **[2]**

79 Paramagnetism is a property shown by transition metal ions or complexes which have at least one unpaired electron. Such transition metal ions or complexes are attracted into a magnetic field with a force which increases as the number of unpaired electrons increases.

I/P

For the elements scandium to zinc give the symbols of:

(a) two +2 ions having two unpaired electrons **[2]**

(b) the +2 ion having most unpaired electrons **[1]**

(c) the +1 ion having no unpaired electrons **[1]**

(d) the +3 ion having no unpaired electrons. **[1]**

80 A number of isomers of dimethylhexene are shown (**A–I**). Select pairs to illustrate each of the following classifications of isomers.

(a) structural:

    (i) position of side chain on straight chain

    (ii) open chain or ring

    (iii) positional of side chain on a ring

    (iv) functional group position on chain. **[4]**

O   **(b)** *cis–trans* spatial (only one isomer is illustrated) **[1]**

    (c) optical **[1]**

    (d) conformation of ring **[1]**

$$\textbf{A}\quad CH_3-CH_2-CH_2-\underset{\underset{\displaystyle CH_3}{|}}{\overset{\overset{\displaystyle CH_3}{|}}{C}}-CH=CH_2$$

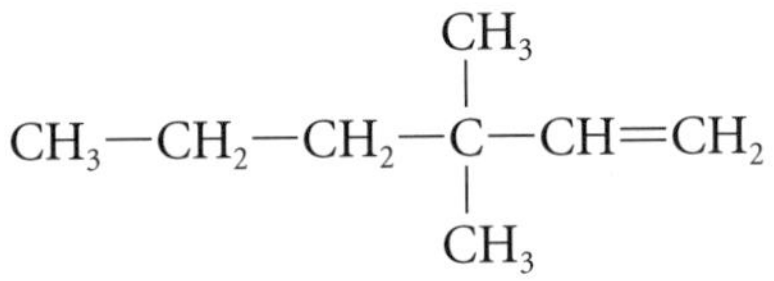

$$\textbf{B}\quad CH_3-CH_2-\underset{\underset{\displaystyle CH_3}{|}}{\overset{\overset{\displaystyle CH_3}{|}}{C}}-CH_2-CH=CH_2$$

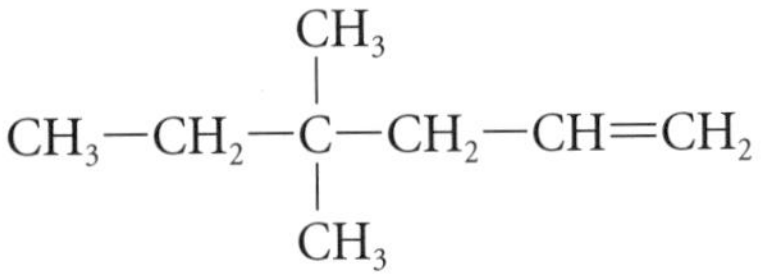

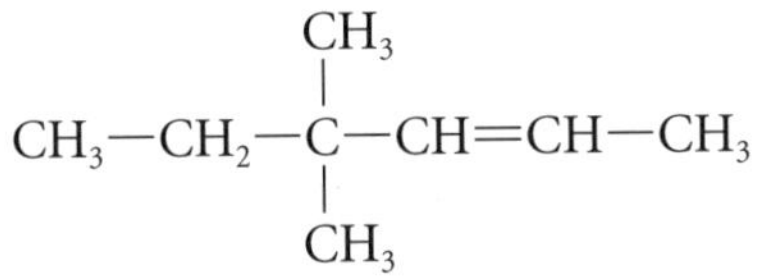

$$\textbf{C}\quad CH_3-CH_2-\underset{\underset{\displaystyle CH_3}{|}}{\overset{\overset{\displaystyle CH_3}{|}}{C}}-CH=CH-CH_3$$

$$\textbf{D}\quad CH_3-\underset{\underset{\displaystyle CH_3}{|}}{CH}-\underset{\underset{\displaystyle CH_3}{|}}{CH}-CH_2-CH=CH_2$$

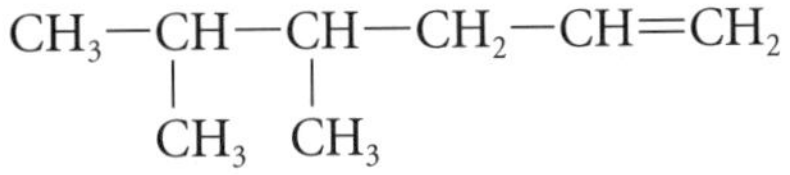

E

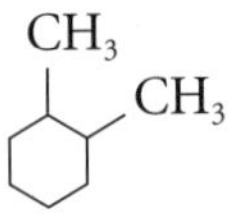

F

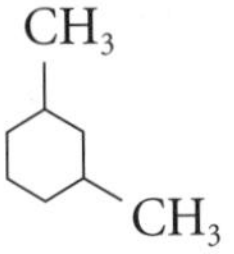

G

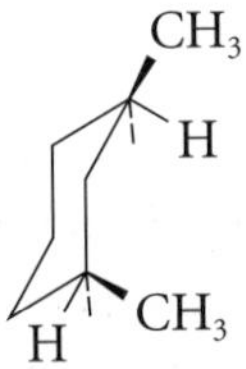

H

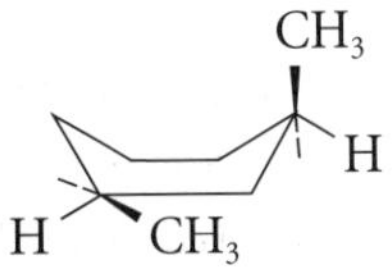

I

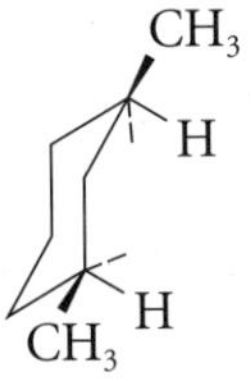

81 Each of the six C—C bonds in benzene is the same length.

    (a) Explain why the Kekulé structures for benzene can be misleading. **[2]**

    (b) Draw the orbitals and the bonding system for benzene. **[2]**

O/P   **(c)** For phenylamine show how the lone pair is involved in the bonding system. **[1]**

    (d) Use the Kekulé structures and curly arrows to show how the lone pair on the nitrogen is able to enhance the electron density at three carbons in the benzene ring. **[3]**

82* **(a)** Write the formulae of the products formed when HCN is added to each of the following:
  (i) CH_3COCH_3
  (ii) CH_3CH_2CHO
  (iii) 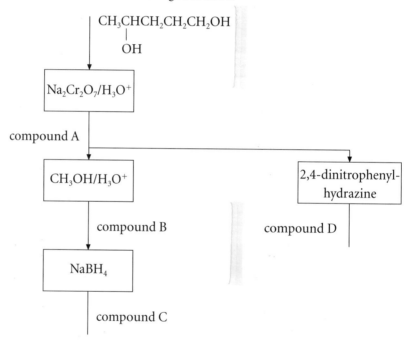CHO
  (iv) $\bigcirc$COCH$_3$ **[4]**

O **(b)** Lithium tetrahydridoaluminate(III) was added to each of the products obtained in part **(a)**. Write the formulae and name each of the amines which are formed. **[8]**

(c) Lithium tetrahydridoaluminate(III) was added to each of the starting materials shown in part **(a)**. Write the formulae and name each of the products formed. **[8]**

83 Complete the flow chart. In your answer write the compound or reactant letter and the formula of the missing substance.

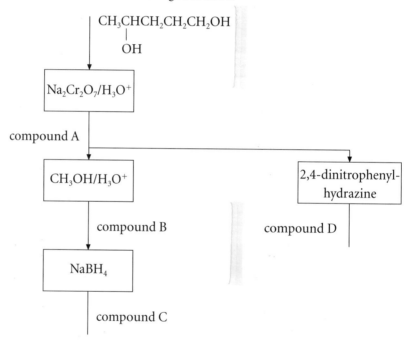

$$CH_3CHCH_2CH_2CH_2OH$$
$$|$$
$$OH$$

$Na_2Cr_2O_7/H_3O^+$

compound A

CH_3OH/H_3O^+

2,4-dinitrophenyl-hydrazine

compound B

compound D

$NaBH_4$

compound C

[4]

84 Complete the flow chart. In your answer write the compound or reactant letter and the formula of the missing substance.

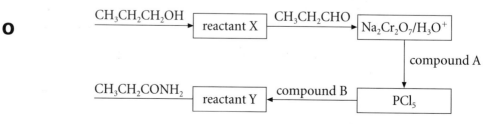

$CH_3CH_2CH_2OH$ → reactant X → CH_3CH_2CHO → $Na_2Cr_2O_7/H_3O^+$

compound A

$CH_3CH_2CONH_2$ → reactant Y ← compound B ← PCl_5

[4]

85 In your answer write the compound or reactant letter and the formula of the missing substance.

O

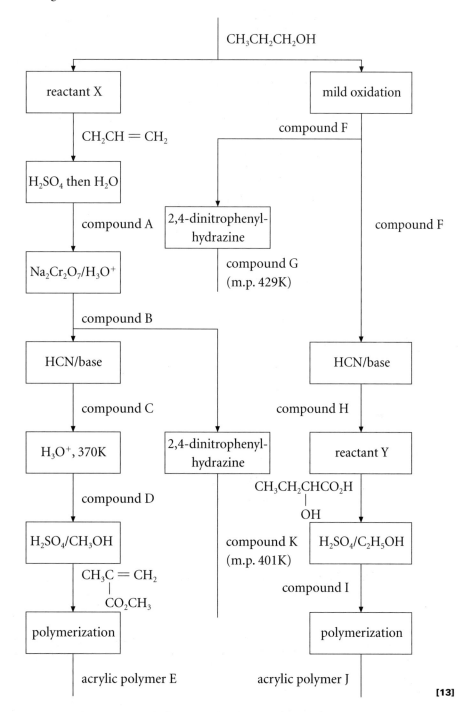

$$CH_3CH_2CH_2OH$$

reactant X   mild oxidation

$$CH_2CH = CH_2$$

compound F

$$H_2SO_4 \text{ then } H_2O$$

compound A   2,4-dinitrophenyl-hydrazine   compound F

$$Na_2Cr_2O_7/H_3O^+$$

compound G (m.p. 429K)

compound B

HCN/base   HCN/base

compound C   compound H

$$H_3O^+, 370K$$ 2,4-dinitrophenyl-hydrazine   reactant Y

$$CH_3CH_2CHCO_2H$$
$$|$$
$$OH$$

compound D

$$H_2SO_4/CH_3OH$$ compound K (m.p. 401K) $$H_2SO_4/C_2H_5OH$$

$$CH_3C = CH_2$$
$$|$$
$$CO_2CH_3$$

compound I

polymerization   polymerization

acrylic polymer E   acrylic polymer J

[13]

86 **(a)** Use the information below to identify compound D, which has the molecular formula $C_8H_8O_2$.

mass spectrum	peaks at m/e 136 and a very intense peak at m/e 91
IR spectrum	3570, 3082, 3040 and strong absorptions at 1780 and 1110 cm^{-1}
NMR spectrum	11.8δ (singlet, 1 proton), 7.1δ (singlet, 5 protons), 3.6δ (singlet, 2 protons)/ppm **[10]**

O/A

(b) Write down the structural formula and name of the organic product formed when compound D reacts with each of the following:

  (i) methanol in the presence of a little concentrated sulphuric acid under reflux

  (ii) phosphorus pentachloride at room temperature

  (iii) lithium aluminium hydride in ethoxyethane under reflux. **[6]**

(c) How might infra-red spectroscopy help in following the progress of the reaction in **(b)(iii)**? **[1]**

Part 4 Answers

Part 1 AS questions

1 Formulae, equations and amount of substances

1.1 The unit of 'amount of substance' is the *mole*. **[1]**

1.2 One *mole* of any substance is the amount of substance which contains as many elementary entities as there are carbon atoms **[1]** in exactly 12 grams **[1]** of pure carbon-12. **[1]** [The nature of the elementary entities should always be specified clearly and may be atoms, molecules, ions, electrons etc.]

1.3 The Avogadro constant is the constant of proportionality **[1]** between the amount of substance (in mol) and the number of specified entities of that substance. **[1]** It is, therefore, the number of specified entities **[1]** per mole. **[1]**
[2 marks max. for any form of answer which embraces the main points]

1.4 (a) The relative atomic mass of an element is the mass of one atom of the element **[1]** on a scale chosen so that the mass of one atom of the ^{12}C isotope of carbon **[1]** is exactly 12 units. **[1]**

(b) The relative formula mass of a substance is the sum of the relative atomic masses **[1]** of all the atoms **[1]** making up the formula of the substance. **[1]**
Neither of the quantities relative atomic mass or relative formula mass has units. **[1]**

(c) The molar mass of a substance is the mass of one mole of the substance. **[1]** It has units of g mol^{-1}. **[1]**

1.5 (a) 20, **(b)** 2, **(c)** 32, **(d)** 256, **(e)** 128, **(f)** 52, **(g)** 126, **(h)** 59, **(i)** 134, **(j)** 248, **(k)** 294, **(l)** 193, **(m)** 199, **(n)** 499, **(o)** 324.
(a)–(j) inclusive **[1 mark each]; (k)–(o)** inclusive **[2 marks each]**. No units should be given – remove one mark if they are.

1.6 (a) 16 g, **(b)** 32 g, **(c)** 64 g, **(d)** 122 g, **(e)** 170 g, **(f)** 60 g, **(g)** 342 g, **(h)** 286 g, **(i)** 370.5 g, **(j)** 478 g.
(a)–(f) inclusive **[1 mark each]; (g)–(j)** inclusive **[2 marks each]**. Units should be given – remove one mark if they are not but ignore the odd omission.

1.7 (a) 2.8 g, **(b)** 70 g, **(c)** 0.044 g, **(d)** 16 g, **(e)** 90 g, **(f)** 11.6 g, **(g)** 776 g, **(h)** 2.46 g, **(i)** 124.5 g, **(j)** 0.57 g.
(a)–(h) inclusive **[1 mark each]; (i) [2 marks], (j) [3 marks]**. Units should be given – remove one mark if they are not but ignore the odd omission.

1.8 (a) 156 g, **(b)** 0.52 g, **(c)** 57.2 g.
[2 marks for each part]. Units should be given – remove one mark if the units are not included at least twice.

1.9 (a) 0.20, **(b)** 0.20, **(c)** 0.10, **(d)** 0.032, **(e)** 0.050, **(f)** 0.050, **(g)** 0.80, **(h)** 20, **(i)** 0.060, **(j)** 0.018.
(a)–(f) inclusive **[1 mark each], (g)–(j)** inclusive **[2 marks each]**.

1.10 (a) 0.010, **(b)** 0.010, **(c)** 0.020, **(d)** 20, **(e)** 0.050, **(f)** 0.050, **(g)** 0.000050, **(h)** 0.0040, **(i)** 0.000083.
(a)–(i) inclusive, **[1 mark each], (j)** volume of room = 96 m^3 **[1]** = 96 000 dm^3 **[1]**, therefore no. of moles of molecules = 4000. **[1]**

1.11 (a) 2.4 dm^3, **(b)** 24 dm^3, **(c)** 48 dm^3, **(d)** 4.8 dm^3, **(e)** 0.24 cm^3, **(f)** 9600 dm^3, **(g)** 9 600 000 dm^3.
(a)–(g) inclusive **[1 mark each]**,
(h) 0.01 mol of CO **[1]** + 0.04 mol of CO_2 **[1]** = 0.05 mol in total ∴ volume = 1.2 dm^3 **[1]**
(i) no. of mol of air = 3.0 **[1]** therefore, no. of mol of oxygen molecules = 0.60 **[1]**, ∴ volume = 14.4 dm^3 **[1]**
(j) 1 mol of methane molecules gives 1 mol of gaseous carbon dioxide molecules under normal laboratory conditions **[1]**; (ignore volume of liquid water) 0.0192 g of methane = 0.0012 mol of methane molecules, therefore 0.0012 mol of gaseous molecules formed under normal laboratory conditions. **[1]** ∴ volume = 28.8 cm^3 **[1]**

1.12 (a) ZnO, **(b)** K_2S, **(c)** $AlCl_3$, **(d)** NH_4NO_2, **(e)** CuBr, **(f)** Li_3PO_4, **(g)** $Ba(OH)_2$, **(h)** Na_2SO_4, **(i)** $Ca(ClO)_2$, **(j)** $Ca(ClO_3)_2$, **(k)** $KClO_4$, **(l)** $Mg(HCO_3)_2$, **(m)** $Fe_2(SO_4)_3$,

(n) $Pb(NO_3)_2$, **(o)** NaI, **(p)** AgF,

(q) $Cr(NO_3)_3$, **(r)** $Sr(NO_3)_2$, **(s)** Al_2S_3,

(t) Li_2CO_3, **(u)** $Mg_3(PO_4)_2$, **(v)** Cu_2O,

(w) CuO, **(x)** FeI_3, **(y)** $(NH_4)_3PO_4$,

(z) Mg_3N_2, **(aa)** $CaSO_3$, **(ab)** $MgSO_4$,

(ac) K_3N, **(ad)** $Al(OH)_3$ **[1 mark each.]**

1.13 (a) 0.4 g of hydrogen = 0.4 moles of hydrogen atoms, 6.4 g of oxygen = 0.4 moles of oxygen atoms. **[1]** Therefore molar ratio of $H : O = 1 : 1$, therefore empirical formula is HO. **[1]**

(b) $FeBr_2$. **[2]**

(c) Mass of sulphur combined with aluminium = $(1.50 \text{ g} - 0.54 \text{ g}) = 0.96$ g, 0.54 g of aluminium = 0.02 moles of aluminium atoms, 0.96 g of sulphur = 0.03 moles of sulphur atoms **[1]**,

Therefore molar ratio of $Al : S = 1 : 1.5$ i.e. $2 : 3$, therefore empirical formula is Al_2S_3 **[1]**

(d) P_2O_5 **[2]** **(e)** COS **[2]** **(f)** C_3H_4 **[2]**

(g) $CaSiO_3$ **[2]** **(h)** $CaCO_3$ **[2]** **(i)** CH_2O **[2]**

(j) $FeSO_4.7H_2O$ **[2]**

1.14 (a) C_6H_6 **(b)** C_5H_{10} **(c)** P_4O_6 **(d)** CON_2H_4

(e) Hg_2Cl_2 **[3 marks each.]**

1.15 (a) $2Mg + O_2 \rightarrow 2MgO$

(b) $CaCO_3 \rightarrow CaO + CO_2$

(c) $2H_2 + O_2 \rightarrow 2H_2O$

(d) $Mg + 2HCl \rightarrow MgCl_2 + H_2$

(e) $2SO_2 + O_2 \rightarrow 2SO_3$

(f) $2Fe + 3Br_2 \rightarrow 2FeBr_3$

(g) $CH_4 + 2O_2 \rightarrow CO_2 + 2H_2O$

(h) $C_3H_7OH + 4\frac{1}{2}O_2 \rightarrow 3CO_2 + 4H_2O$ (or doubled)

(i) $2Ag_2CO_3 \rightarrow 4Ag + 2CO_2 + O_2$

(j) $Fe_2O_3 + 3H_2 \rightarrow 2Fe + 3H_2O$

(k) $CaCO_3 + 2HCl \rightarrow CaCl_2 + H_2O + CO_2$

(l) $4CuO + CH_4 \rightarrow 4Cu + CO_2 + 2H_2O$

(m) $3CuO + 2NH_3 \rightarrow 3Cu + N_2 + 3H_2O$

(n) $4NH_3 + 5O_2 \rightarrow 4NO + 6H_2O$

(o) $2KOH + H_2SO_4 \rightarrow K_2SO_4 + H_2O$

(p) $Ba(NO_3)_2 + Na_2SO_4 \rightarrow BaSO_4 + 2NaNO_3$

(q) $2HNO_3 + Mg(OH)_2 \rightarrow Mg(NO_3)_2 + 2H_2O$

(r) $2KHCO_3 \rightarrow K_2CO_3 + H_2O + CO_2$

(s) $Pb(NO_3)_2 + 2KI \rightarrow PbI_2 + 2KNO_3$

(t) $Cu + 4HNO_3 \rightarrow Cu(NO_3)_2 + 2NO_2 + 2H_2O$

[3 marks for each fully correct equation.]

1.16 (a) $C_3H_4 + 4O_2 \rightarrow 3CO_2 + 2H_2O$ (or doubled)

(b) $C_3H_6 + H_2 \rightarrow C_3H_8$

(c) $C_{10}H_{22} \rightarrow 2C_2H_4 + C_6H_{14}$

(d) $C_3H_8 + 2Cl_2 \rightarrow C_3H_6Cl_2 + 2HCl$

(e) $C_3H_6(OH)_2 + 2Na \rightarrow C_3H_6(ONa)_2 + H_2$

(f) $C_4H_6 + 2Br_2 \rightarrow C_4H_6Br_4$

(g) $MgBr_2 + 2AgNO_3 \rightarrow Mg(NO_3)_2 + 2AgBr$

(h) $KCl + H_2SO_4 \rightarrow KHSO_4 + HCl$

or $2KCl + H_2SO_4 \rightarrow K_2SO_4 + 2HCl$

(i) $2HI + H_2SO_4 \rightarrow SO_2 + 2H_2O + I_2$

or $6HI + H_2SO_4 \rightarrow S + 4H_2O + 3I_2$

or $8HI + H_2SO_4 \rightarrow H_2S + 4H_2O + 4I_2$

(j) $KClO_4 \rightarrow KCl + 2O_2$

(k) $Al(OH)_3 + 3HNO_3 \rightarrow Al(NO_3)_3 + 3H_2O$

(l) $PbO_2 + 4HCl \rightarrow PbCl_2 + 2H_2O + Cl_2$

(m) $Li_2CO_3 \rightarrow Li_2O + CO_2$

(n) $(NH_4)_2Cr_2O_7 \rightarrow Cr_2O_3 + N_2 + 4H_2O$

(o) $(CH_3CO_2)_2Pb + K_2CrO_4 \rightarrow PbCrO_4 + 2CH_3CO_2K$

[3 marks for each fully correct equation.]

1.17 (a) $H^+(aq) + OH^-(aq) \rightarrow H_2O(l)$

(b) $Cu(s) + 2Ag^+(aq) \rightarrow Cu^{2+}(aq) + 2Ag(s)$

(c) $2S_2O_3^{2-}(aq) + I_2(aq) \rightarrow S_4O_6^{2-}(aq) + 2I^-(aq)$

(d) $CO_3^{2-}(aq) + 2H^+(aq) \rightarrow CO_2(g) + H_2O(l)$

or $CO_3^{2-}(aq) + H^+(aq) \rightarrow HCO_3^-(aq)$

(e) $Pb^{2+}(aq) + 2I^-(aq) \rightarrow PbI_2(s)$

(f) $2Fe^{2+}(aq) + 2H^+(aq) + H_2O_2(aq) \rightarrow 2Fe^{3+}(aq) + 2H_2O(l)$

(g) $2CrO_4^{2-}(aq) + 2H^+(aq) \rightarrow Cr_2O_7^{2-}(aq) + H_2O(l)$

(h) $2ClO_2(g) + 2OH^-(aq) \rightarrow 2ClO_2^-(aq) + \frac{1}{2}O_2(g) + H_2O(l)$, or doubled

(i) $2Cr^{3+}(aq) + 2Zn(s) + 2H^+(aq) \rightarrow 2Cr^{2+}(aq) + 2Zn^{2+}(aq) + H_2(g)$, or halved

(j) $4MnO_4^-(aq) + 4OH^-(aq) \rightarrow 4MnO_4^{2-}(aq) + O_2(g) + 2H_2O(l)$

[3 marks for each fully correct equation.]

1.18 (a) Balanced equation is: $CaCO_3 \rightarrow CaO + CO_2$, **[1]**

molar mass of $CaCO_3 = 100$ g mol^{-1}

molar mass of $CaO = 56$ g mol^{-1}, **[1]**

from equation 1 mol CaO is produced from 1 mol $CaCO_3$, **[1]**

therefore 56 g CaO is produced from 100 g $CaCO_3$,

therefore 56 t CaO is produced from 100 t $CaCO_3$,

therefore 500 t CaO is produced from $500/56 \times 100$ t $CaCO_3 = 893$ t $CaCO_3$. **[1]**

(b) Balanced equation is:

$C_6H_{12}O_6 \rightarrow 2C_2H_5OH + 2CO_2$, **[1]**

mass of glucose = 22.5 g. **[3]**

(c) Equation is:

$Pb(NO_3)_2 + Mg \rightarrow Pb + Mg(NO_3)_2$ **[1]**,

mass = 0.83 g. **[3]**

(d) Equation is: $Mg + 2H^+ \rightarrow Mg^{2+} + H_2$ **[1]**

mass = 0.48 g. **[3]**

(e) Balanced equation is:

$TiCl_4 + 4Na \rightarrow Ti + 4NaCl$ **[1]**,

mass of titanium = 1.20 t; mass of

sodium = 2.30 t. **[3]**

1.19 **(a)** 70.0% **[2]**

(b) (i) 6.0 cm³ of ethanoic anhydride = 6.48 g

= 0.064 mol, **[1]** 4.0 cm³ of phenylamine

= 4.08 g = 0.044 mol, therefore

phenylamine is the limiting reagent. **[1]**

(ii) 0.044 mol of phenylamine produce

0.044 mol of N-phenylethanamide

= 5.9 g, **[1]** therefore percentage

yield = 3.8/5.9 × 100% = 64%. **[1]**

(c) 19.1 g. **[2]**

1.20 **(a)** 25.0/1000 × 0.1 mol = 0.0025 mol,

(b) 3.75 mol, **(c)** 0.006 mol, **(d)** 0.0034 mol,

(e) 0.0056 mol, **(f)** 1.00 mol, **(g)** 0.0002 mol,

(h) 0.000054 mol, **(i)** 0.14 mol,

(j) 0.000515 mol. **[1 mark for each part.]**

1.21 **(a)** 1.0 M, **(b)** 0.10 M, **(c)** 0.25 M,

(d) 0.0033 M, **(e)** 0.050 M, **(f)** 0.50 M,

(g) 0.12 M, **(h)** 0.10 M, **(i)** 0.050 M,

(j) 0.11 M. **[2 marks for each part.]**

1.22 **(a)** 0.0025 mol, **(b)** 0.030 mol, **(c)** 0.045 mol,

(d) 0.090 mol, **(e)** 0.25 mol.

[1 mark for each part.]

1.23 **(a)** Molar mass of sodium hydroxide

= 40 g mol⁻¹ **[1]**

therefore mass required

= 100/1000 × 0.10 mol × 40 g mol⁻¹

= 0.40 g **[1]**

(b) 1.70 g **[2]** **(c)** 5.80 g **[2]** **(d)** 0.069 g **[2]**

(e) 24.95 g **[2]**

1.24 **(a)** $KOH + HCl \rightarrow KCl + H_2O$ **[1]**

(b) 0.00135 mol **[1]**

(c) 0.00135 mol **[1]**

(d) 0.135 mol dm⁻³ **[1]**

1.25 0.57 mol dm⁻³ **[1]**

1.26 0.021 mol dm⁻³ **[3]**

1.27 **(a)** methyl orange, bromophenol blue etc., NOT

phenolphthalein **[1]**

(b) $Na_2CO_3 + 2HCl \rightarrow 2NaCl + CO_2 + H_2O$ **[1]**

(c) 0.00197 mol **[1]** **(d)** 0.000984 mol **[1]**

(e) 0.0984 mol dm⁻³ **[1]**

(f) 28.1 g **[1]** **(g)** 96% **[1]**

2 States of matter and changes of state

2.1

property	solids	liquids	gases
arrangement of particles	regular	random	random
distance apart of particles	very close together – touching	close together	far apart
movement of particles	vibrate	move randomly and quite slowly	move randomly and quickly
forces between particles	strong	strong	weak
diffusion	little	slow	rapid
volume	fixed	fixed	variable, they expand to fill the whole volume of the container
shape	fixed	variable, they take the shape of the part of the container they occupy	variable, they take the shape of the container
density	high	high	low
effect of pressure	cannot be compressed	can be compressed very slightly	easily compressed

[Expect at least 6 properties; 1 mark for each property to a maximum of 6.]

2.2 Any appropriate suggestions. **[1 mark for each.]**

2.3 Brownian motion is the movement of small particles as a result of molecular bombardment. **[1]** In observing both the smoke cell and the pollen grains, random movement is seen but the Brownian motion is much faster in the smoke cell. **[2]** Bombardment of the smoke particles is due to air molecules which move much faster and with more energy than the water molecules which bombard the pollen grains. Moreover, pollen grains are much larger than smoke particles and so require more energy to move them. **[3]**

2.4 **(a)** Particles in a liquid are much closer together than in a gas so the forces between them are greater and therefore a liquid is much more resistant to compression. **[2]**

(b) Pressure is caused by the force exerted by molecules colliding with the walls of their container. At higher temperatures they hit the walls more frequently and with greater energy, therefore they exert a greater force. **[2]**

(c) Because the movement of the gas particles is fast and random, gases soon spread out to fill the whole of any container into which they are put. **[2]**

(d) A liquid boils when the pressure exerted by the escaping molecules is the same as the surrounding pressure. At the top of Mont Blanc, the pressure is much lower than at sea level so the liquid does not have to be heated so much for the pressure exerted by the escaping molecules to be equal to that of the ambient pressure. **[2]**

2.5 The pressure will double. **[1]** When the plunger is pushed in, the particles will hit the walls of the container more frequently because its volume is smaller. However, they do not change their average speed so they hit the walls with the same force as before. **[2]**

2.6 Correctly drawn diagram **[3]**

(a)

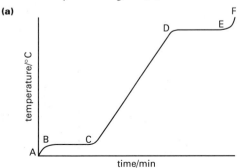

(b) The diagram should have five distinct areas: A–B; temperature rising as more energy is given to the ice particles which therefore vibrate more rapidly, **[2]** B–C; temperature steady as energy is being used to loosen the molecules rather than to raise the temperature – the ice is melting, **[2]** C–D; all the ice has now melted and the temperature is rising as energy is given to the particles which therefore move around more rapidly – some will even escape from the liquid, **[2]** D–E; temperature steady as energy is being given to the liquid particles to overcome the remaining forces of attraction between them and to separate them completely, **[2]** E–F; all the liquid has now boiled and the temperature is rising as energy is being used to make the particles move around more quickly. **[2]**

2.7 Carbon dioxide has a molecular structure; although there are strong covalent bonds within the molecules there are only weak forces between molecules so that they are easily separated from each other and the boiling point is therefore low. **[4]**

Silicon dioxide has a giant structure; all the atoms in the lattice are joined by strong forces (strong covalent bonds) and so they can only be separated from each other with considerable difficulty. Thus much energy is required to boil this substance and its boiling point is high. **[4]**

2.8 **copper** – metallic bonding, strong forces between atoms (or discussion of forces between positive ions and surrounding 'sea of electrons'), therefore high melting and boiling points (strong forces still largely retained in the liquid state); mobile electrons in the 'sea of electrons' allow electricity to be conducted,

diamond – giant lattice of strong covalent bonds, very strong forces between atoms, therefore high melting and boiling points; all electrons used in bonding, therefore electricity cannot be conducted,

iodine – molecular structure, relatively weak forces between molecules, therefore low melting and boiling points; no free electrons therefore does not conduct,

lithium chloride – giant ionic lattice, strong forces of attraction between ions, (strong forces still largely retained in the liquid state), therefore high melting and boiling points; ions not free to move in the solid state but when melted, they can do so and this allows electricity to be conducted as the ions move towards the oppositely charged electrodes.

phenol – as for iodine.

[3 marks each.]

2.9 **A** – metal, atoms closely packed together in an ordered arrangement, therefore high melting and boiling point; metallic bonding – conduction by free electrons.

B – molecular substance, only weak forces between molecules, therefore low melting and boiling point; no free electrons or ions to make conduction of electricity possible.

C – giant ionic lattice, strong forces of attraction between ions, therefore high melting and boiling point; ions not free to move until melted or dissolved in water, then conduction of electricity is possible.

D – molecular substance, only weak forces between molecules, therefore low melting and boiling points; no ions therefore does not conduct when solid or when melted but does form ions by reaction with water so conducts in aqueous solution.

E – as for **B**.

F – probably a giant covalent lattice, high melting and boiling points and no conduction of electricity in solid or liquid states; therefore no free electrons or ions.

G – metal, as for **A**, but note that substance is a liquid at room temperature: must be mercury as it is the only liquid metal at room temperature; difficult to explain why this is so!

H – giant ionic lattice, as for **C**; has particularly strong forces of attraction between ions, hence very high melting and boiling points and insolubility in water.

[3 marks each.]

3 Ionic bonding

3.1 **(a)** An atom is electrically neutral but an ion carries a positive or negative charge. **[1]**

(b) (i) Metals. **[1]**

(ii) Non-metals. **[1]**

(c) An electrostatic force of attraction between particles with opposite charges. **[1]**

(d) Electrons. **[1]**

(e) Electrons are lost or gained in order that the ion formed has the electron configuration of the nearest noble gas. Thus, for example, an atom of a Group 2 element, which has two electrons in its outermost main energy level, loses the two electrons to form an ion of charge $+2$, the electron configuration of the ion formed being the same as that of the preceding noble gas. **[1]** for initial statement, **[1]** for suitable example.

3.2 **(a)** 2, 8, 1. **[1]** **(b)** 2, 8. **[1]**

(c) 2, 8, 7. **[1]** **(d)** 2, 8, 8. **[1]**

(e) Electron arrangement of atoms

Electron arrangement of ions.

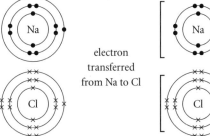

electron transferred from Na to Cl

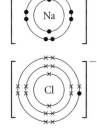

[2]

(f) Alternate sodium and chloride ions **[1]** arranged in interpenetrating face-centred cubes, **[1]** each cube being formed of either all sodium or all chloride ions. **[1]** Alternatively a diagram may be drawn but for full marks, the diagram must have some three-dimensional character. **[3]**

3.3 **(a)** Li^+, **(b)** 10, **(c)** 31, **(d)** 25, **(e)** 24, **(f)** 47, **(g)** 35, **(h)** O^{2-}, **(i)** P^{3-}, **(j)** 54.

[1 mark each.]

3.4 **(a)** Ca^{2+}, **(b)** Al^{3+}, **(c)** S^{2-}, **(d)** Cu^+, **(e)** Cu^{2+}, **(f)** N^{3-}, **(g)** H^+, **(h)** H^-, **(i)** F^-, **(j)** K^+. **[1 mark each.]**

3.5

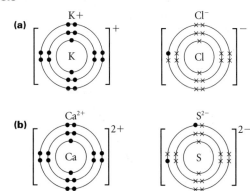

(a)

(b)

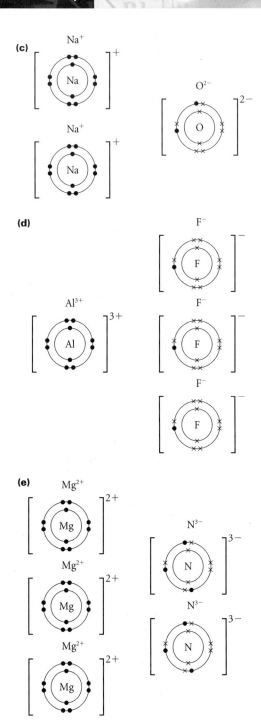

(c)

(d)

(e)

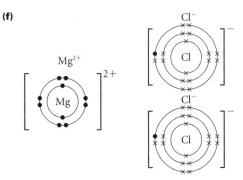

(f)

[2 marks each.]

3.6 (i) Ionic bonds are strong, therefore a solid requires a considerable amount of energy to loosen the ions in the lattice structure and so to melt the solid. Strong forces of attraction between ions of opposite charge are still retained in the liquid state and so boiling points also tend to be very high.

(ii) Water is a polar molecule. The energy given out when an ion–dipole interaction forms between the charged ends of water molecules and oppositely-charged ions, leading to solvation of the ions, compensates for the energy required to break down the ionic lattice and so ionic compounds are commonly soluble in water.

(iii) Many organic solvents are non-polar or only slightly polar. They, therefore, cannot form strong ion–dipole interactions, as in the case of water, and so insufficient energy is released to compensate for the energy required to break down the lattice and so ionic compounds are usually insoluble in organic solvents.

(iv) Ions are not free to move in the solid state but when an ionic substance is melted or dissolved in water, the ions become free to move and so are attracted to oppositely charged electrodes. As the ions move towards the electrodes, they enable electricity to be conducted through the liquid.

[4 marks each.]

3.7 (a) NO_3^-, **(b)** NO_2^-, **(c)** PO_4^{3-}, **(d)** MnO_4^-, **(e)** ClO_3^-, **(f)** BrO^-, **(g)** $Cr_2O_7^{2-}$, **(h)** SO_4^{2-}, **(i)** MnO_4^{2-}, **(j)** IO_4^-. **[1 mark each.]**

3.8 (a) $FeBr_2$, **(b)** $FePO_4$, **(c)** $ZnSO_4$, **(d)** $Ba(MnO_4)_2$, **(e)** Ag_2CrO_4, **(f)** $KBrO$, **(g)** $KBrO_3$, **(h)** $Al(NO_3)_3$, **(i)** $NaNO_2$, **(j)** Na_2SO_3. **[1 mark each.]**

3.9 There are a number of possible experiments. Two possible choices are:

(i) Placing a crystal of potassium manganate(VII) in the centre of a damp piece of filter paper cut to fit a microscope slide and connecting a d.c. supply of about 12–20 V to the two ends; a purple streak will be seen to move towards the positive connection as the manganate(VII) ions are attracted towards it.

(ii) Electrodes are placed in a U-tube containing dilute hydrochloric acid placed above a solution of copper(II) chromate(VI). Using a 12–20 V d.c. supply, the colours of the copper(II) and chromate(VI) ions are soon seen moving towards the oppositely charged electrodes. **[3 marks each.]**

3.10 (a) Students should realize from the description that magnesium oxide has the same type of structure as sodium chloride and should draw a clear diagram of alternately arranged Mg^{2+} and O^{2-} ions. For full marks the diagram should have some three-dimensional character and not be simply the face of a cube. To show the interpenetrating face-centred cubic aspect for both ions, it should have at least four, and preferably five, ions in each edge. **[4]**

(b) (i) A unit cell is the smallest part of a crystal structure which is characteristic of the whole crystal. **[1]**

(ii)

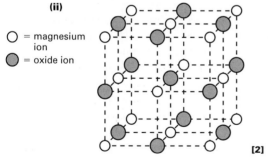

○ = magnesium ion
⬤ = oxide ion

[2]

(c) (i) Co-ordination number is the number of nearest neighbours that a particular ion has in a crystal lattice. **[2]**

(ii) Both are 6 – a magnesium ion is surrounded by six oxide ions and an oxide ion is surrounded by six magnesium ions. **[2]**

(d) One would expect magnesium oxide to have considerably higher melting and boiling points than sodium chloride, as the charges on both ions are bigger and the forces of attraction between them will therefore be considerably larger. **[3]**

(e) e.g. another alkali metal halide such as potassium bromide. **[1]**

3.11 (a) The diagram should show two interpenetrating cubes: a face-centred cubic arrangement of calcium ions and a simple cubic arrangement of fluoride ions. **[4]**

(b) Each calcium ion is surrounded by eight fluoride ions, therefore 8. Each fluoride ion is surrounded by four chlorine atoms, therefore 4. **[2]**

3.12 (a) Ions are not 'hard', rigid spheres like snooker balls but 'soft', more like rubber balls. Such behaviour would be expected of a diffuse cloud of electrons. Ions can, therefore be distorted from their spherical nature by surrounding ions of opposite charge. This leads to a degree of sharing of electrons between ions. **[3]**

(b) Ion polarization. **[1]**

(c) No 'ionic' bond is purely ionic. There is always some degree of covalent character within an ionic substance. The ions are therefore somewhat distorted from a purely spherical shape and the energy holding the ions together within the lattice will change slightly. It will also affect the extent of solubility in both polar and non-polar solvents etc. **[2 marks for 2 sensible suggestions.]**

(d) The larger the charge on the positive ion, the more it is likely to distort a neighbouring negative ion.
The smaller the positive ion is, the more the charge is concentrated upon it, and so the greater will be the distortion of a surrounding negative ion.
The larger the charge on the negative ion, the more likely it is to be distorted by surrounding positive ions.
The larger the negative ion is, the less concentrated will be its charge and so the more easily distorted it will be. **[4]**

(e) There will be much greater ion polarization in beryllium iodide than in barium fluoride **[1]** as the beryllium ion would be much smaller than the barium ion and the iodide ion would

be much larger than the fluoride ion. **[1]**
Therefore a greater degree of covalent
character in the bonding of beryllium
iodide. **[1]**

4 Energetics

4.1 The term 'enthalpy' (or heat content) refers to the
total energy content of a system held at constant
pressure. **[1]** (Its absolute value cannot be
determined.) The term 'enthalpy change' refers to
the heat given out or taken in when substances
react. **[1]** (It is the difference between the total
enthalpy of the products and the total enthalpy of
the reactants.)

4.2 ΔH stands for enthalpy change; **[1]** the $\pm$ sign
indicates whether the heat is being taken in or
given out during the change occurring in the
system, a $+$ sign implying an endothermic
reaction (heat absorbed, so that the system has a
greater heat content than before the change) **[1]**
and a $-$ sign an exothermic reaction (heat
evolved, so that the system has a lower heat
content than before the change). **[1]**
xkJ mol^{-1} refers to the size of the enthalpy change
for the quantities specified in the equation, the
units being those of an energy quantity. **[1]**

4.3 The standard conditions for measuring enthalpy
changes are:
(i) one atmosphere pressure,
(ii) a temperature of 298 K (25 °C),
(iii) substances in their normal physical state at
1 atmosphere and the stated temperature,
(iv) for solutions, 1.0 M solutions. **[4]**

4.4 Diagrams of the type shown below:

(i)

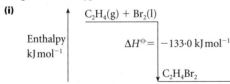

(ii)

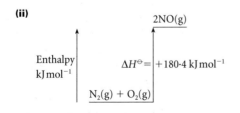

**[2 marks for each – it is not necessary for them
to be drawn to scale.]**

4.5 (a) The standard enthalpy change of formation
of a compound is the enthalpy change that
takes place when one mole of the compound
is formed from its constituent elements
under standard conditions. **[3]**
(b) The standard enthalpy change of combustion
of a substance is the enthalpy change that
occurs when one mole of the substance
undergoes complete combustion under
standard conditions. **[3]**

4.6 (a) Hess's law states that the total enthalpy
change accompanying a chemical change is
independent of the route by which the
chemical change takes place. **[2]** (It is a form
of the principle of conservation of energy
applied to chemical systems.)
(b) (i) $\Delta H^{\ominus} = +178.3$ kJ mol^{-1} **[2]**
(ii) $\Delta H^{\ominus} = -851.5$ kJ mol^{-1} **[2]**
(iii) $\Delta H^{\ominus} = -74.8$ kJ mol^{-1} **[2]**
The specific name of the enthalpy change
calculated is the 'standard enthalpy change of
formation of methane'. **[2]**

4.7 (a) Correct Hess cycle leading to verification of
answer. **[4]**
(b) Correct Hess cycle leading to
$\Delta H^{\ominus} = -2720.9$ kJ mol^{-1}. **[4]**
The specific name of the enthalpy change
calculated is the 'standard enthalpy change of
combustion of cyclobutane'. **[2]**

4.8 Enthalpy change = 8360 J or 8.36 kJ. **[2]**

4.9 (a) For calcium + water = 2571 J, **[2]**
calcium hydroxide + hydrochloric
acid = 730.3 J, **[2]**
calcium + hydrochloric acid = 3031 J **[2]**
(b) For calcium + water, $\Delta H = -321.3$ kJ mol^{-1}, **[2]**
calcium hydroxide + hydrochloric acid,
$\Delta H = -91.3$ kJ mol^{-1}, **[2]**
calcium + hydrochloric acid,
$\Delta H = -418.0$ kJ mol^{-1}. **[2]**
(c) Ca(s) + 2H$_2$O(l) → Ca(OH)$_2$(aq) + H$_2$(g), **[2]**
Ca(OH)$_2$(aq) + 2HCl(aq) → CaCl$_2$(aq) +
2H$_2$O(aq), **[2]**
Ca(s) + 2HCl(aq) → CaCl$_2$(aq) + H$_2$(g). **[2]**
(d) Correct Hess cycle. **[2]**
(e) One route -412.7 kJ mol^{-1}, other route
-417.9 kJ mol^{-1}, therefore good agreement
within the limits of experimental error (heat
loss is the main source of error in
thermochemical experiments). **[4]**

4.10 (a) (i) C_3H_8 **[1]**

(ii) $C_3H_8 + 5O_2 \rightarrow 3CO_2 + 4H_2O$ **[2]**

(b) -50.4 kJ g^{-1} **[1]**

(c) 50 400 kJ kg^{-1} **[1]**

4.11 (a) Successively -654, -653, -654 (all values in kJ mol^{-1}). **[2]**

(b) Virtually the same. Each time a —CH$_2$ group is added, so a specific quantity of energy appears to be linked to specific bonds. **[3]**

(c) -3967 kJ mol^{-1} + $(-654$ kJ mol$^{-1})$ = 4621 kJ mol^{-1}. (Alternatively draw a graph of enthalpy change of combustion against the number of carbon atoms in a molecule and extrapolate to required value.) **[2]**

(d) $C_7H_{15}OH + 10\frac{1}{2}O_2 \rightarrow 7CO_2 + 8H_2O$ (or doubled) **[2]**

Correct energy level diagram. **[2]**

4.12 (a) $C_3H_6(g) + 4\frac{1}{2}O_2(g) \rightarrow 3CO_2(g) + 3H_2O(l)$ **[2]**

(b) (i)

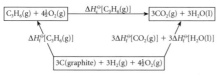

(ii) Standard enthalpy change of formation of propene = $+20.2$ kJ mol^{-1} **[2]**

(c) (i) Standard enthalpy change of formation of propan-1-ol = -302.7 kJ mol^{-1} **[2]**

(ii) Standard enthalpy change of formation of phenol = -165.0 kJ mol^{-1} **[2]**

4.13 The equation for the complete combustion of phosphine is:

$4PH_3(g) + 8O_2(g) \rightarrow P_4O_{10}(s) + 6H_2O(l)$

Using this equation together with the correct cycle and data, a value of -1180.1 kJ mol^{-1} is obtained for the standard enthalpy change of combustion of phosphine. **[5]**

4.14 (a) (i) Bonds present in buta-1,2-diene = 6 C—H bonds, 2 C=C bonds and 1 C—C bond.

therefore enthalpy change of atomization = $6(+413) + 2(+612) + 347$ kJ mol^{-1} = $+4049$ kJ mol^{-1} **[3]**

(ii) Realization that calculation is identical to that in **(a) (i)**, **[2]** if calculation is fully reworked, then only. **[1]**

(b) Cycle:

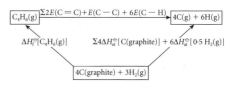

Calculation leads to values of $+4012.5$ kJ mol^{-1} for buta-1,2-diene and $+4064.9$ kJ mol^{-1} for buta-1,3-diene. **[4]**

(c) The difference in values between the enthalpy changes of atomization of buta-1,2-diene and buta-1,3-diene as calculated in **(a)** and **(b)** is due to the fact that average bond energies were used in the calculation in **(a) [1]**; in fact bond energies vary slightly with the environment of the bonds. **[1]**

5 Kinetics

5.1 (a) The term 'rate of reaction' refers to how quickly a particular substance is used up in a chemical reaction **[1]** or how quickly a product is formed. **[1]** (If a reaction takes place in solution, it is likely to refer to the rate of change of concentration of a particular substance.)

(b) Obviously answers are going to vary widely – in any case there is no exact definition of what we mean by a very fast reaction. However, a likely choice would be a reaction such as the neutralization of hydrogen ions by hydroxide ions or a reaction in which there is a simple electron transfer between atoms. Another possible reaction would be the break-up of halogen molecules by intense light energy of an appropriate frequency and their subsequent recombination. Such reactions are complete in a very small fraction of a second. **[2]**

(c) As for **5.1 (b)**, there is no definition of what is meant by a very slow reaction. Some pupils might consider the ripening of a fruit to be a very slow reaction; others the erosion of a limestone building or tombstone by acid rain or the decay of a long-lived radioactive isotope. An interesting example is the conversion of one isomer of aspartic acid into another in fossil bones at ordinary temperatures. Half of the sample of aspartic acid is converted in about 100 000 years; this is probably the slowest known reaction. **[2]**

5.2 (a) Choose **four** from: concentration of the reactants, pressure (for a gaseous reactant), temperature of the reactants, the presence of a catalyst, surface area (for a solid reactant), and the influence of light (in photochemical reactions). **[4]**

(b) The answers given depend on the selection made in **5.2 (a)**, but are likely to be drawn from:
- increased concentration – increase in the rate of reaction,
- increased pressure (for a gaseous reactant) – as for increased concentration,
- increased temperature – increases the rate of reaction,
- the presence of a catalyst – increases the rate of reaction,
- increased surface area (for a solid) – increases the rate of reaction.
- influence of light – some reactions go very much faster when a bright light is shone upon the reactants. **[4]**

5.3 (a) For a reaction to occur, particles of the reactants must collide with each other. **[1]**

(b) • increased concentration – the reacting particles are more likely to meet and react,
- increased pressure – for gaseous reactants, as for increased concentration,
- increased temperature – the particles move faster and have more energy, they therefore meet more often and are more likely to react when they do so,
- the presence of a catalyst – makes a reaction take place more easily, for example, by bringing gaseous reactants closer together on the surface of a solid catalyst,
- increased surface area – smaller 'bits' of a solid have a larger surface area exposing more of the solid to attack by another reactant,
- influence of light – may result in the production of more reactive species and so are more likely to react when they meet.

[2 marks for each factor discussed.]

5.4 (a) Chemical bonds in the reactants must be broken **[1]** and new bonds made to enable the products to be formed. **[1]**

Bond breaking involves the absorption of energy (endothermic change). **[1]**
Bond making involves the release of energy (exothermic change). **[1]**
Do make certain that you get these changes the right way round – students very commonly fail to do so.

(b) When particles collide they do not have sufficient energy to enable chemical bonds in the reactants to be broken. **[1]**
Particles must collide in the right way – (a 'glancing blow' is very different to a 'head on' collision). **[1]**

(c) By introducing the idea of 'activation energy' **[1]** – the minimum energy which must be possessed by reacting particles if they are to react successfully. **[1]**
By introducing a 'steric factor' **[1]** to allow for the fact that particles have to be in a suitable orientation to react. **[1]**

(d) At a higher temperature, many more particles collide with an energy equal to or greater than the activation energy, **[1]** so a much higher proportion of collisions is successful in bringing about chemical change (or a much higher proportion of collisions result in chemical bonds being broken). **[1]**

5.5 (a) A 'catalyst' is a substance which increases the rate of a chemical reaction **[1]** without itself becoming permanently involved in the reaction. **[1]**

(b) One would expect at least **four** features to be discussed. They are likely to be selected from the following:
- A small amount of a catalyst will produce a large amount of chemical change.
- There is the same mass of catalyst present at the end of a chemical reaction as there was at the beginning. This does not mean that the catalyst has not been involved in some way in the reaction. A catalyst may change its physical state as a result of its involvement.
- Catalysts are often very specific to a particular chemical reaction.
- A catalyst does not affect the amount of product(s) obtained from specific amounts of reactants, merely the rate at which the product(s) is/are obtained.

- The surface area of a catalyst can often considerably affect its efficiency.
- The efficiency of a catalyst is often increased or decreased considerably by traces of other substances (promoters).
  [2 marks for each feature.]

(c) A catalyst provides an alternative route for a reaction **[1]** with a lower activation energy. **[1]** A higher proportion of collisions between reacting particles will therefore be successful, **[1]** thus increasing the rate of reaction. **[1]**

5.6 (a) The tungsten is acting as a catalyst. **[1]** The activation energy for the reaction is lower in the presence of tungsten **[1]** enabling the reaction to take place more quickly. **[1]**

(b) A suitable energy profile would be:

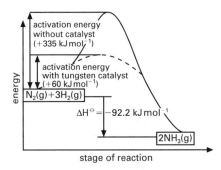

stage of reaction

It is not necessary to draw the diagram to scale but it should show clearly labelled axes, **[1]** what is meant by activation energy, **[1]** the correct values of activation energy for the catalysed **[1]** and uncatalysed reactions **[1]** and indicate that the reaction concerned is endothermic (i.e. the products should have a higher energy than the reactants). **[1]**

5.7 (a) A 'homogeneous' catalyst is one which is in the **same** physical state as the reactants involved in the reaction. **[1]**
A 'heterogeneous' catalyst is one which is in a **different** physical state to the reactants involved in the reaction. **[1]**

(b) Any correct reactions given should score **[1]** mark each, with a maximum of two for homogeneous reactions and two for heterogeneous reactions. **[4]**

5.8 (a) **Similarities** of enzymes and conventional catalysts:
- both increase the rate of a reaction but do not affect it in any other way,

- both enable a reaction to take place by an alternative route of lower activation energy,
- both can be very easily 'poisoned' by small amounts of particular impurities,
- both are only needed in small quantities.

(b) **Differences** between enzymes and conventional catalysts:
- enzymes are much more efficient catalysts than conventional catalysts,
- enzymes are more highly specific than many conventional catalysts,
- enzyme activity increases with temperature until an optimum temperature is reached, after which it rapidly decreases,
- enzyme activity is affected by pH – different enzymes work best at different pH values.

[Look for at least six features in total for a maximum mark of 6]. They are likely but not exclusively to be taken from the lists above.

6 Equilibrium

6.1 (a) A 'reversible reaction' is one which can be made to take place in either direction; reactants can be converted to products and products can be converted to reactants. **[1]**

(b) A state of 'equilibrium' would be reached. The amounts of each of the substances present in the mixture would stay constant. **[2]**

6.2 (a) • An equilibrium state can only be reached in a closed system; one in which matter cannot enter or leave.
- The equilibrium state can be reached from either direction, that is, in relation to a chemical equation, from either the reactants or the products.
- Equilibrium is a dynamic state. Although there are no changes in concentration of substances in the mixture, the reactions are still continuing in both directions but at the same rate.
- An equilibrium system, although stable under a particular set of conditions, is sensitive to alteration in those conditions.
  [4 marks for a sensible set of statements such as those above.]

(b) The amount of potassium iodide dissolved would stay the same, as the solution is saturated at that temperature. **[1]**

(c) The solution would become radioactive. **[1]**
The observation indicates that the equilibrium state is dynamic, since the increase in radioactivity of the solution implies that some of the added labelled potassium iodide has dissolved, but since the solution is already saturated, an equal amount of potassium iodide must have crystallized, indicating that changes in both ways are continuing. **[2]**

6.3 (a) The amounts of A, B, C and D will be constant. **[1]**

(b) They will be the same. **[1]**

(c) They will be numerically equal, but of opposite sign, that is, the reaction in one direction will be exothermic and in the other direction endothermic. **[2]**

(d) The amounts of each substance present at equilibrium would be the same (therefore no effect on the position of equilibrium) but equilibrium would be reached more quickly.
[2]

6.4 (a) (i) Increased temperature results in less product. **[1]**

(ii) The reaction is exothermic (ΔH is negative). **[1]**

(b) (i) Increased pressure favours the formation of product. **[1]**

(ii) There is a reduction in the number of gas molecules present ($4 \rightarrow 2$). **[1]**

(c) Le Chatelier concluded that when a change is made to a system at equilibrium, the reaction tries to oppose the change being made. **[1]**
Thus, in **(a)** temperature increased, therefore the endothermic change is favoured as this absorbs heat, **[1]** therefore reaction moves in favour of the reactants. **[1]**
In **(b)** increased pressure will be opposed by a decrease in the number of molecules present, **[1]** this is achieved by the formation of more product. **[1]**

(d) They are optimum conditions. **[1]**
A higher pressure will be more expensive – greater running costs etc.
A lower pressure will give a lower yield.
A higher temperature will give a lower yield.
A lower temperature will give a better yield but a slower rate.
[1 mark for each correct point made to a maximum of 4.]

6.5 (a) (i) Less products – increased pressure favours the formation of fewer molecules, therefore equilibrium shifts to the left.

(ii) Less products as **(i)**.

(iii) No change – same number of molecules on both sides of the equation.

(iv) More products – increased pressure favours the formation of fewer molecules, therefore equilibrium shifts to the right. **[2 marks each.]**

(b) (i) fewer product molecules,

(ii) more product molecules,

(iii) very little difference in amounts of products,

(iv) more product molecules. **[1 mark each.]**

7 The structure of the atom

7.1 (a) (i) neutron **[1] (ii)** electron. **[1]**

(b) proton and neutron. **[2]**

(c) in (spherical) shells/orbits around the nucleus held there by electrostatic attraction. **[2]**

7.2 (a) (i) The number of protons in the nucleus of an atom. **[1]**

(ii) It is same as the number of electrons in an atom. **[1]**

(b) He 2p 2e, Na 11p 11e, Na^+ 11p 10e, O 8p 8e, O^{2-} 8p 10e, F 9p 9e, F^- 9p 10e, Al 13p 13e, Al^{3+} 13p 10e, N 7p 7e, N^{3-} 7p 10e. **[11]**

7.3 (a) Because a proton and an electron have the same (but opposite) charges and the atom overall must be neutral. **[2]**

(b) In the atom the numbers of protons and electrons are equal, removing an electron to form an ion leaves the atom with one more proton, hence an overall positive charge. **[2]**

(c) The total number of protons and neutrons in the nucleus of an atom. **[1]**

(d) The number of protons in an atom of that element. **[1]**

7.4 (a) Isotopes **[1]**

(b) $^{11}_{5}B$ $^{10}_{5}B$ **[2]**

(c) 10.8 **[2]**

(d) abundance

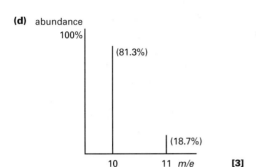

(81.3%)

(18.7%)

10   11 *m/e* **[3]**

7.5 (a) 24.33 **[2]**

(b) abundance

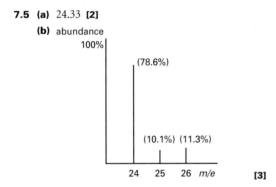

(78.6%)

(10.1%) (11.3%)

24 25 26 *m/e* **[3]**

7.6 (a) Ne, **(b)** 10, **(c)** 22, **(d)** 10, **(e)** 7, **(f)** 3,
(g) 2, **(h)** 4, **(i)** +1, **(j)** Al^{3+}, **(k)** 13, **(l)** 10,
(m) 14. **[13]**

7.7 (a) H^-, **(b)** 1, **(c)** 1, **(d)** −1, **(e)** Cl^-,
(f) 35, **(g)** 17, **(h)** −1, **(i)** 18, **(j)** 8, **(k)** 10,
(l) 10, **(m)** −2. **[13]**

7.8 (a) The RAM must be greater than 28 because all
isotopes have relative isotopic masses greater
than this but cannot be as high as 28.5
because over 90% of the atoms have a relative
isotopic mass of 28. **[2]**

(b) They all have the same atomic number of 14.
[1]

(c) Number of neutrons = mass no. − atomic
number,
i.e. 28 − 14 = 14, 29 − 14 = 15,
30 − 14 = 16 neutrons. **[2]**

8 The electronic configuration of atoms

8.1 (a) The one furthest away from it. **[1]**

(b) 1st = 2, 2nd = 8, 3rd = 18. **[3 × 1]**

8.2 (a)

[5 × 1]

(b) All the energy levels in the neon atom are full.
[1]

(c) Neon is unreactive, because full electron
shells are stable. **[2]**

8.3 (a) K, N, Cl, Al **[4 × 1]**

(b) Ti^{2+}, Mn^{3+} **[2 × 1]**

8.4 Maximum number of electrons; s = 2, p = 6,
d = 10. **[3 × 1]**

8.5 N = $1s^2\ 2s^2\ 2p^3$, Na = $1s^2\ 2s^2\ 2p^6\ 3s^1$, Ne = $1s^2\ 2s^2$
$2p^6$, S = $1s^2\ 2s^2\ 2p^6\ 3s^5\ 3p^4$, Ca = $1s^2\ 2s^2\ 2p^6\ 3s^2$
$3p^6\ 4s^2$. **[5 × 1]**

8.6 (a) 5 = $1s^2\ 2s^2\ 2p^1$, 8 = $1s^2\ 2s^2\ 2p^4$, 12 = $1s^2\ 2s^2$
$2p^6\ 3s^2$, 9 = $1s^2\ 2s^2\ 2p^5$, 14 = $1s^2\ 2s^2\ 2p^6\ 3s^2$
$3p^2$, 18 = $1s^2\ 2s^2\ 2p^6\ 3s^2\ 3p^6$. **[6 × 1]**

(b) Li^+ = $1s^2$, S^{2-} = $1s^2\ 2s^2\ 2p^6\ 3s^2\ 3p^6$, Cl^- = $1s^2$
$2s^2\ 2p^6\ 3s^2\ 3p^6$, Al^{3+} = $1s^2\ 2s^2\ 2p^6$, Ca^{2+} = $1s^2$
$2s^2\ 2p^6\ 3s^2\ 3p^6$, K^+ = $1s^2\ 2s^2\ 2p^6\ 3s^2\ 3p^6$. **[6 × 1]**

8.7 (a) (i) 2 = Li^+, 10 = Na^+, Mg^{2+} or Al^{3+},
18 = K^+, Ca^{2+} or Sc^{3+}. **[3 × 1]**

(ii) 2 = H^-, 10 = N^{3-}, O^{2-} or F^-,
18 = P^{3-}, S^{2-} or Cl^-. **[3 × 1]**

(b) Ne = Na^+, Mg^{2+} or Al^{3+}, He = Li^+,
F^- = Na^+, Mg^{2+} or Al^{3+}, S^{2-} = K^+, Ca^{2+} or
Sc^{3+}, Ar = K^+, Ca^{2+} or Sc^{3+}, C^{4-} = Na^+,
Mg^{2+} or Al^{3+}. **[6 × 1]**

8.8 Na^+ = N^{3-}, O^{2-} or F^-, Mg^{2+} = N^{3-}, O^{2-} or F^-,
Ar = P^{3-}, S^{2-} or Cl^-, K^+ = P^{3-}, S^{2-} or Cl^-,
He = H^-. **[5 × 1]**

8.9 argon = p, calcium = s, carbon = p, iron = d,
magnesium = s, scandium = d, sodium = s,
vanadium = d. **[8 × 1]**

9 Electronegativity and its application to band type

9.1 Electronegativity is a measure of the ability of an atom in a molecule to attract electrons to itself in a chemical bond. **[2]**

9.2 (a) (i) Cl and N, or C and S **(ii)** B and F **[2]**
　　(b) (i) F and O **[2]**
　　　(ii) hydrogen bonds **[1]**
　　　(iii) Two from: boiling point, viscosity, density, solubility, heats of vaporization and melting point. **[2]**
　　(c) $B^{\delta+}-H^{\delta-}$, $C^{\delta+}-O^{\delta-}$, $Cl^{\delta+}-F^{\delta-}$, $S^{\delta-}-H^{\delta+}$, $C^{\delta+}-F^{\delta-}$, $N^{\delta-}-H^{\delta+}$ **[6]**

9.3 Electronegativity decreases with increasing atomic number down a group because the outer electron shells are further from the nucleus **[1]** and shielded from it by more intervening shells of electrons. **[1]**
Electronegativity increases with increasing atomic number across a period because the nuclear charge increases **[1]** and the electron added at the same time is not as strongly shielded from that increased nuclear charge. **[1]**

9.4 CF_4 – tetrahedral, **[1]** not polar, **[1]**
NH_3 – pyramidal, **[1]** CO_2 – linear, **[1]** not polar, **[1]** H_2O – angular, **[1]** SO_2 – angular, **[1]**
CH_3Cl – tetrahedral, **[1]** $BeCl_2$ – linear, **[1]** not polar, **[1]** XeF_4 – square planar, **[1]** not polar. **[1]**

9.5 Ethanal < ethanol < ethanoic acid. **[1]**
Ethanoic acid forms a dimer **[1]** via hydrogen bonds **[1]** between the O—H and the C=O group, **[1]** ethanol has hydrogen bonds but does not form a dimer, **[1]** whereas ethanal has only (weaker) dipole–dipole forces. **[1]**

9.6 Ethanol, CH_3CH_2OH (78 °C) has intermolecular hydrogen bonds, **[1]** which are stronger **[1]** than the van der Waals' and dipole–dipole intermolecular forces in methoxymethane, CH_3OCH_3 (−23 °C). **[1]**

9.7

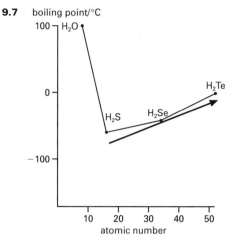

Water has an unusually high boiling point because of intermolecular hydrogen bonding. **[2]** The increase in boiling point from H_2S to H_2Te arises because intermolecular van der Waals' forces increase in magnitude as the number of electrons in the atoms increases. **[2]**
axes **[2]** points on graphs **[2]**.

9.8 (a) Carbon dioxide is a simple molecular substance **[1]** in which the intermolecular forces are weak van der Waals' forces, **[1]** while silicon dioxide is a macromolecular solid. **[1]** This difference arises from the marked reluctance of silicon to form multiple bonds, (due to the poorer overlap of the larger 3p orbitals). Rather than form two double bonds (like carbon in CO_2), 4 single Si—O bonds are energetically favoured and a giant structure results. **[1]**

(b) In graphite, only three of the electrons of each carbon atom are used in bonding, forming layers of carbon atoms consisting of hexagons. **[1]** Between these layers only weak van der Waals' forces exist, **[1]** so they slide easily over one another and so graphite is soft. The electrons not used in forming covalent bonds are free to move throughout the structure and hence graphite conducts electricity well. **[1]** In the diamond structure all 4 electrons are used in bonding, forming a giant molecular structure. All the bonds are very strong so with no points of weakness diamond is very hard. There are no free electrons to conduct electricity. **[1]**

10 The factors influencing the shapes of molecules and ions

10.1 CH_4 BP 4 LP 0, NH_3 BP 3 LP 1
$BeCl_2$ BP 2 LP 0, H_2O BP 2 LP 2
BF_3 BP 3 LP 0, SF_4 BP 4 LP 1. **[12]**

10.2 **(a)** NH_4^+ BP 4 LP 0, BH_4^- BP 4 LP 0,
CH_3^+ BP 3 LP 0, PCl_4^+ BP 4 LP 0,
PCl_6^- BP 6 LP 0, ICl_4^- BP 4 LP 2. **[12]**

(b) NH_4^+, BH_4^- and PCl_4^+ tetrahedral, CH_3^+ trigonal planar, PCl_6^- octahedral, ICl_4^- square planar **[6]**

10.3 **(a)** 2 lone pairs repel most **[1]**

(b) 2 bond pairs repel least **[1]**.
Lone pairs of electrons tend to be closer to the central nucleus than bond pairs which are pulled out from the nucleus by the other bonding atom. Lone pairs therefore repel more than bond pairs since their electron density is more concentrated. **[4]**

10.4

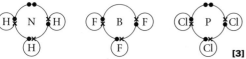

Only BF_3 has bond angles of 120° as three bond pairs distribute themselves equally and as far away from each other as possible. The other two molecules have a lone pair. Four pairs of electrons distribute themselves tetrahedrally but since a lone pair repels more than a bond pair, the molecules are pyramidal with bond angles less than the tetrahedral bond angle of 109.5°. **[2]**

10.5

CH_3^+ trigonal planar

ICl_2^+ angular

H_3O^+ and ClO_3 pyramidal

[6]

10.6 The molecules are isoelectronic, i.e. have the same number of electrons in total. **[1]** Adding an electron to B gives it the same number of electrons as C, while removing an electron from N has the same effect. **[1]** The molecules BH_4^-, CH_4 and NH_4^+ must therefore have the same number of bond and lone pairs, hence the same shape. **[1]** NH_3 has the same shape as H_3O^+, because removing an electron from oxygen (giving O^+) gives it the same electron

configuration as N **[1]** and since NH_3 and H_3O^+ have the same number of bonds, they have the same shape. **[1]**

10.7 CH_4 is symmetrical with 4 BP **[1]** but NH_3 has a LP in place of one of the BP which repels the remaining BP to a greater extent, **[1]** reducing the bond angle. **[1]** In H_2O two BP have been replaced by LP so, extending the previous argument the bond angle will be reduced even further. **[1]**

10.8 Both have the same number of BP (3) and LP (1), **[1]** but the lone pair on the P atom will be larger than that on N **[1]** and appears to exert a greater repulsion on the three BP, resulting in a reduction in the bond angle. **[1]**

10.9 Silicon has vacant d-orbitals (unlike carbon) **[1]** if nitrogen forms an sp^2 hybrid, a lone pair remains in a 2p orbital **[1]** which can overlap with the vacant 3d orbital in silicon to form a pi-bond **[1]** which increases the N—Si bond strength **[1]** and this overlap is a maximum when the molecule is planar. **[1]**

10.10 Xe has 8 electrons in its valence shell **[1]** and 6 electrons required for the Xe—F bonds. **[1]** This leaves a lone pair in the XeF_6 molecule **[1]** therefore distorting the expected perfectly octahedral shape. **[1]**

11 The concept of oxidation state and its uses

11.1 **(a)** O_2 zero, O^{2-} −2, O_2^{2-} −1 **[3]**

(b) CO +2, CO_2 +4, MgO +2, Al_2O_3 +3, B_2O_3 +3, Cl_2O_7 +7, SO_3 +6, P_4O_{10} +5. **[8]**

11.2 **(a)** HF (+1), OF_2 (+2), BF_3 (+3), CF_4 (+4), ClF_5 (+5), SF_6 (+6), IF_7 (+7). **[7]**

(b) N_2 zero, NH_3 −3, N_2H_4 −2, NH_4^+ −3, NF_3 +3, N^{3-} −3. **[6]**

(c) N_2O +1, NO +2, N_2O_3 +3, N_2O_4 and NO_2 +4, N_2O_5 +5. **[6]**

11.3 SO_3^{2-} +4, SO_4^{2-} +6, $S_2O_3^{2-}$ +2, $S_4O_6^{2-}$ +2.5. **[4]**

11.4 **(a)** +1, **(b)** +2, **(c)** +3, **(d)** +2. **[4]**

11.5 reductions
$N_2 \rightarrow NH_3$ (−3)
$SO_4^{2-} \rightarrow SO_3^{2-}$ (−2)
$SO_3 \rightarrow SO_3^{2-}$ (−2)
oxidations
$NH_3 \rightarrow NO_3^-$ (+8)
$N_2H_4 \rightarrow N_2$ (+2)
$H_2S \rightarrow SO_2$ (+6)

$H_2O \rightarrow H_2O_2$ (+1)

$S_2O_3^{2-} \rightarrow SO_4^{2-}$ (+4)

no change

$NO_2 \rightarrow N_2O_4$

$NH_3 \rightarrow NH_4^+$

$SO_2 \rightarrow SO_3^{2-}$

$H_2O \rightarrow H_3O^+$

$12 \times$ **[1]** for each correct classification

$8 \times$ **[1]** for each correct change in oxidation state.

11.6 (a) 2, (b) 2, (c) 6, (d) 2. **[4]**

11.7 (a) $1e^-$, (b) $2e^-$, (c) $4e^-$, (d) $2e^-$. **[4]**

11.8 (a) $SO_4^{2-} + 8H^+ + 6e^- \rightarrow S + 4H_2O$ **[2]**

(b) $NO_3^- + 9H^+ + 8e^- \rightarrow NH_3 + 3H_2O$ **[2]**

(c) $BrO_3^- + 6H^+ + 5e^- \rightarrow \frac{1}{2}Br_2 + 3H_2O$ **[2]**

(d) $2ClO^- + 4H^+ + 2e^- \rightarrow Cl_2 + 2H_2O$ **[2]**

11.9 (a) $SO_4^{2-} + 4H^+ + 2e^- \rightarrow H_2SO_3 + H_2O$ **[2]**

(b) $NO_3^- + 4H^+ + 3e^- \rightarrow NO + 2H_2O$ **[2]**

(c) $N_2 + 8H^+ + 6e^- \rightarrow 2NH_4^+$ **[2]**

(d) $H_2O_2 + 2H^+ + 2e^- \rightarrow 2H_2O$ **[2]**

11.10 The following species can, in theory, undergo disproportionation: N_2, CO, H_2O_2 **[4 × 1]**. The remainder represent either the highest OS of the elements which they contain $[SO_4^{2-}, NO_3^-]$ (hence cannot be further oxidized) **[1]** or represent the lowest OS of the element which they contain $[NH_3 NH_4^+Cl^- S^{2-}]$ (hence cannot be further reduced). **[1]**

11.11 $BrO_3^- \rightarrow Br^-$ and BrO_4^- and $SO_3^{2-} \rightarrow S$ and SO_4^{2-} **[2]**

11.12 (a) $3IO^- \rightarrow 2I^- + IO_3^-$ **[3]**

(b) $4ClO_3^- \rightarrow Cl^- + 3ClO_4^-$ **[3]**

(c) $3NO \rightarrow N_2O + NO_2$ **[3]**

(d) $3NO_2^- + 2H^+ \rightarrow NO_3^- + 2NO + H_2O$ **[3]**

(e) $3S + 2H_2O \rightarrow SO_2 + 2S^{2-} + 4H^+$ **[3]**

(f) $7N_2O + 9H_2O \rightarrow 6NH_3 + 8NO_2$ **[3]**

11.13 Since the oxidizing agent (NO_2^-) and the species oxidized (NH_4^+) are present in the salt in a $1 : 1$ mole ratio, **[1]** the change in oxidation state undergone by each must be the same, **[1]** in order that the electrons should balance. **[1]** This implies that the oxidation state of nitrogen in the product must be the mean of the oxidation states of nitrogen in NO_2^- and NH_4^+, **[1]** i.e. $(+3 + (-3)) \div 2 = 0$. **[1]** The product obtained is therefore nitrogen, so $NH_4NO_2 \rightarrow N_2 + 2H_2O$ **[1]**

11.14 Since the oxidizing agent (NO_3^-) and the species oxidized (NH_4^+) are present in the salt in a $1 : 1$ mole ratio, **[1]** the oxidation state of nitrogen in the product must be the mean of the oxidation states of nitrogen in NO_3^- and NH_4^+, **[1]** i.e.

$[5 + (-3)] \div 2 = +1$ **[1]** and so the product in this case is nitrous oxide, N_2O. **[1]**

$NH_4NO_3 \rightarrow N_2O + 2H_2O$ **[2]**

11.15 (a) 25.0 cm^3

(b) 75.0 cm^3

(c) 12.5 cm^3. **[3 × 3]**

11.16 (a) 50.0 cm^3

(b) 16.0 cm^3

(c) 48.0 cm^3

(d) 15.0 cm^3. **[4 × 3]**

11.17 (a) 0.17 M

(b) 0.083 M

(c) 0.033 M. **[3 × 3]**

11.18 (a) (i) 0.195×10^{-3} mol

(ii) 0.975×10^{-3} mol **[3]**

(b) 0.055 g **[2]**

(c) 1.37 g **[2]**

(d) 91.0% **[2]**

11.19 (a) 2.06×10^{-3} mol **[2]**

(b) 1.03×10^{-3} mol **[2]**

(c) 1.03×10^{-2} mol **[2]**

(d) 1.03×10^{-2} mol **[2]**

(e) 0.247 dm^3 **[2]**

(f) 0.412% **[2]**

11.20 (a) 2.0×10^{-3} mol **[1]**

(b) 10^{-2} mol **[2]**

(c) 2×10^{-2} mol **[1]**

(d) $n = 5$ **[3]**

12 Atomic structure and the Periodic Table

12.1 H $1s^1$, Na $1s^2 2s^2 2p^6 3s^1$, Al $1s^2 2s^2 2p^6 3s^2 3p^1$, N $1s^2 2s^2 2p^3$, Ca $1s^2 2s^2 2p^6 3s^2 3p^6 4s^2$, Cl $1s^2 2s^2 2p^6 3s^2 3p^5$. **[6]**

12.2 Li$^+$ $1s^2$, Mg^{2+} $1s^2 2s^2 2p^6$, O^{2-} $1s^2 2s^2 2p^6$, F$^-$ $1s^2 2s^2 2p^6$, P^{3-} $1s^2 2s^2 2p^6 3s^2 3p^6$, Si^{4-} $1s^2 2s^2 2p^6 3s^2 3p^6$. **[6]**

12.3 8 is $1s^2 2s^2 2p^4$, 9 is $1s^2 2s^2 2p^5$, 11 is $1s^2 2s^2 2p^6 3s^1$, 13 is $1s^2 2s^2 2p^6 3s^2 3p^1$, 16 is $1s^2 2s^2 2p^6 3s^2 3p^4$, 17 is $1s^2 2s^2 2p^6 3s^2 3p^5$. **[6]**

12.4 8 is $1s^2 2s^2 2p^6$, 9 is $1s^2 2s^2 2p^6$, 11 is $1s^2 2s^2 2p^6$, 13 is $1s^2 2s^2 2p^6$, 16 is $1s^2 2s^2 2p^6 3s^2 3p^6$, 17 is $1s^2 2s^2 2p^6 3s^2 3p^6$. **[6]**

12.5 N and P, Be and Mg, He and Ar. **[3]**

12.6 4 and 20, 6 and 14, 10 and 18, 9 and 17. Elements in the same group have atomic numbers which differ by 8, or a whole number multiple of 8. **[6]**

12.7 In the modern Periodic Table using atomic numbers each element has one more electron and one more proton than that which precedes it [1] and there are therefore no gaps in the table. [1] The modern form ensures that elements fall naturally into groups within which the outer electronic configuration of the elements are identical and their chemical properties are therefore very similar. [1] The older practice of using relative atomic mass fell victim to the problem of isotopes, [1] which meant that an element could be out of place as a result of changes in its nuclear structure, rather than in its electronic configuration [1] which determines its chemistry. The elements tellurium (Z 52, A_r 127.6) and iodine (Z 53, A_r 126.9) provide an example of elements which would appear in the wrong groups on the basis of their relative atomic masses. [1]

12.8 Ionic radii increase with increasing atomic number as a group is descended [1] and decrease with increasing atomic number as a period is crossed [1]. Magnesium is in the group to the right of lithium, but in the next period [1] the increase due to the group effect is balanced by the period effect and the ionic radii of the two are similar. [1]

13 Reactions of the elements and compounds of Groups 2 and 7

13.1 (a) $Mg^{2+} < Ca^{2+} < Sr^{2+} < Ba^{2+}$ [2]

(b) Atomic radii increase with increase in atomic number [1] because there are more electron shells, which increase the size of the atoms, [1] but also because more shells shield the outer electrons from the attraction of the nucleus. [1]

(c) Barium, [1] because reaction involves electron loss and this is easier for the larger barium atom, [1] because its outer electrons are held least strongly by the nucleus, [1] being further from it [1] and better shielded by intervening electron shells. [1]

13.2 (a) (i) $2Ca(s) + O_2(g) \rightarrow 2CaO(s)$

(ii) $CaO(s) + H_2O(l) \rightarrow Ca(OH)_2(s)/(aq)$

(iii) $Ca(OH)_2(aq) + CO_2(g) \rightarrow CaCO_3(s) + H_2O(l)$ [6]

(b) It may be more energetically unfavourable to form the Mg^{2+} ion, compared to the Ba^{2+} ion (2186 kJ mol^{-1} compared to 1468 kJ mol^{-1}, a difference of 718 kJ mol^{-1}) [2] but the extra

energy required is more than offset by the greater lattice enthalpy of MgO [3889 kJ mol^{-1}] compared to BaO (3152 kJ mol^{-1}), a difference of 734 kJ mol2^{-1}. [2]

13.3 Mg 3600, Ca 2850, Sr 2430, Ba 1920°C. [1] Ionic radii increase from Mg to Ba as more electron shells are added, [1] this increase in size reduces the lattice enthalpy of the oxide [1] and hence its melting point. [1]

13.4 (a) O_2^{2-} [1]

(b) $BaO_2(s) + 2H_2O(l) \rightarrow Ba(OH)_2(aq) + H_2O_2(aq)$ [2]

(c) The peroxide ion is almost certainly larger than the oxide ion [1] and hence its lattice with the small magnesium ion is likely to be less stable than the oxide [1]. Since the other enthalpy terms involving magnesium are constant for both reactions, [1] the peroxide is unstable with respect to the oxide overall. [1]

13.5 (a) $2M(NO_3)_2(s) \rightarrow 2MO(s) + 4NO_2(g) + O_2(g)$ [3]

(b) (i) barium **(ii)** magnesium. [2]

13.6 Solubility decreases with increase in atomic number. [1] Increase in cation radius reduces both the hydration enthalpy of the ions [1] and the lattice enthalpy of the sulphates. [1] The fact that the sulphates become less soluble with increasing cation size implies that the lattice enthalpy is affected less by this change than is the hydration enthalpy. [1]

13.7 The lattice enthalpy of the resulting oxide will be greater than the starting carbonate for the smaller Group 2 ions [1] so the carbonate of the smaller cation decomposes more readily [1] OR the smaller cations polarize (weaken) the C—O bond in the nitrate ion to a greater extent than the larger cations, [1] facilitating its decomposition. [1]

13.8 (a) (i) $4LiNO_3(s) \rightarrow 2Li_2O(s) + 4NO_2(g) + O_2(g)$ [3]

(ii) One of: lithium forms only a simple oxide, unstable peroxide; lithium carbonate decomposes more readily than those of other Group 1 metals; lithium halides show more covalent character than other Group 1 chlorides. [1]

(b) calcium: orange/red, barium: green [2]

(c) (i) The increase in atomic radii caused by adding a further shell of electrons [1] is almost exactly offset by the decrease in atomic radius **[1]** caused by the increased nuclear charge in magnesium, so their radii are similar. **[1]**

(ii) The reaction involves the loss of the outer s-electron **[1]** – this is easier for Cs because the atom is larger **[1]** and the outer 6s electron is considerably more shielded by intervening shells than the outer 3s electron in sodium. **[1]**

(d) 92.7% **[4]**

13.9 The intermolecular bonding is van der Waals' **[1]**. Iodine has more electrons than chlorine **[1]** so the van der Waals' forces are stronger in iodine, hence it is a solid at RT **[1]**.

13.10 (a) A substance/species which readily takes up electrons **[1]** allowing another substance to be oxidised. **[1]**

(b) $Cl_2(g) + 2I^-(aq) \rightarrow 2Cl^-(aq) + I_2(aq)$ **[2]**

(c) Oxidation involves acceptance of an electron **[1]** this process becomes less energetic as the atomic number increases **[1]** because the atoms increase in size/have more intervening electron shells. **[1]**

(d) (i) iodide ion **[1]**

(ii) $Br_2(l) + 2I^-(aq) \rightarrow 2Br^-(aq) + I_2(aq)$ **[2]**

13.11 (a) (i) Concentrated sulphuric acid; room temperature **[2]**

(ii) $NaCl(s) + H_2SO_4(l) \rightarrow NaHSO_4(s) + HCl(g)$ **[2]**

(b) (i) $2Br^- + 2e^- \rightarrow Br_2$
$SO_4^{2-} + 2e^- + 4H^+ \rightarrow SO_2 + 2H_2O$
$2Br^- + SO_4^{2-} + 4H^+ \rightarrow SO_2 + 2H_2O + Br_2$ **[6]**

(ii) Chloride ions are weaker reducing agents than bromide ions **[1]** because they have a smaller ionic radius **[1]** or H_2SO_4 is not a sufficiently strong oxidizing agent to oxidize chloride ions to chlorine and hence give up electrons less easily. **[1]**

13.12 The H—F bond is stronger than the H—Cl bond **[1]** due to the small size of the F atom and so the H—F bond is difficult to break. **[1]** In addition HF is strongly hydrogen bonded. These hydrogen bonds must be broken before individual HF bonds can be broken. **[1]**

13.13 $Cl_2 + 2NaOH(aq) \rightarrow NaCl(aq) + NaClO(aq) + H_2O$ **[2]**

The oxidation state of chlorine has increased from 0 to +1 **[1]** at the same time as it has decreased from 0 to −1. **[1]** Such a change in the oxidation state of an element is described as disproportionation. **[1]**

14 The trends in the physical properties of the elements across a period

14.1 (a) lithium M, carbon GM, nitrogen SM, oxygen SM, fluorine SM, neon A. **[6]**

(b) Carbon is a giant molecular structure so melting involves breaking large numbers of strong **[1]** covalent **[1]** bonds, while nitrogen has only weak (van der Waals') intermolecular forces between its molecules **[1]** and so boils at a lower temperature. **[1]**

14.2 (a) Sodium, magnesium and aluminium all have metallic structures **[1]** in which the conductivity increases as the number of valence electrons increases **[1]**. Silicon, despite having 4 valence electrons, has a lower conductivity because it has the diamond structure **[1]** in which its electrons are largely immobile in covalent bonds. **[1]**

(b) Graphite has a layer structure **[1]** with weak van der Waals' forces between the layers **[1]** making it soft **[1]**, while diamond has a three-dimensional giant structure, which is much more rigid **[1]**. In graphite only three electrons are used per atom in bonding leaving one free to conduct electricity **[1]** but all the electrons in diamond are immobile in covalent bonds. **[1]**

14.3 (a) The general trend is an increase **[1]** caused by increasing nuclear charge **[1]** which causes an increase in the attractive forces between the nucleus and the outer electron **[1]** so making it more difficult to remove.

(b) The 1st IE of boron is lower than that of beryllium because the B electron is lost from the 2p orbital **[1]** which is on average further from the nucleus than the 2s (from which the Be electron comes) **[1]** and hence bound less strongly. **[1]** The electron lost from oxygen is paired in the 2p orbital **[1]** and experiences repulsion from the other electron in that orbital **[1]** and hence requires less energy to ionize it. **[1]**

14.4

	Na	Mg	Al	Si	P	S	Cl
atomic radius/nm	0.191	0.160	0.143	0.117	0.110	0.104	0.099
boiling point/K	1156	1380	2740	2628	553	718	238

[4]

14.5

element (identity)	outer electronic configuration	group
1 (Li)	s^1	1
2 (Mg)	s^2	2
3 (S)	$s^2 p^4$	6
4 (C)	$s^2 p^2$	4

[8]

14.6 (a) plot itself (axes labelled **[2]**, points **[1]**). Gradual increase in boiling point Li–C **[1]** followed by a sharp decrease between carbon and nitrogen **[1]** with no particularly regular trend after nitrogen. **[1]**

(b) (i) The number of valence (bonding) electrons increases from beryllium to carbon **[1]** and there is a trend from metallic to giant covalent which might also increase the boiling point. **[1]**

(ii) Giant covalent **[1]** because the melting points are high **[1]** compared to those for metallic structures with a similar number of bonding electrons. **[1]**

(c) Carbon is macromolecular **[1]** hence much energy is needed to break the interatomic bonds **[1]** while nitrogen (simple molecular) has only weak van der Waals' intermolecular forces. **[1]**

14.7 (a) The first electron in aluminium is lost from the 3p orbital **[1]** which is further from the nucleus than the 3s orbital **[1]** from which magnesium loses its first electron. **[1]**

(b) The second electrons in both aluminium and magnesium are lost from the 3s orbital **[1]** but the loss of this electron in magnesium gives the (stable) neon structure **[1]** and aluminium also has a higher nuclear charge. **[1]**

(c) The third electron in magnesium must be lost from the neon core **[1]** which is particularly stable. **[1]**

14.8 (a) (i) More electrons are involved in the metallic bonding moving across the period from Na to Al **[1]** and the strength of the metallic bond depends directly on the number of electrons involved **[1]** so more energy is needed to boil the metal as you move across **[1]**.

(ii) Greater number of electrons in Al should lead to a significantly greater melting point **[1]** however the structure of Al might involve less-effective packing of the atoms **[1]** so melting point is similar to that of Mg. **[1]** Boiling point [the transition from (l) to (g)] is unaffected by structural differences **[1]** so the effect of the extra electron in Al becomes evident in the higher boiling point. **[1]**

(iii) The van der Waals' forces in the solid and the liquid must be of similar magnitudes. **[1]**

(b) Trends in boiling points are likely to be more reliable **[1]** because the structures of the liquid elements are likely to be similar **[1]** while the melting points will depend on the structure of the solid elements, which can differ considerably. **[1]**

15 Nomenclature and formulae of alkanes and alkenes

15.1 (a) (i) pentane
(ii) 2-methylpentane
(iii) 3-methylpentane
(iv) 2-methylpentane **[4]**

(b) (ii) and **(iv)** **[1]**

(c) (i) C_5H_{12} **[1]**

(d) $CH_3CHCH_2CH_3$ 2-methylbutane **[2]**
    |
    CH_3

    CH_3
    |
CH_3CCH_3 2,2-dimethylpropane **[2]**
    |
    CH_3

15.2 (a) (i) $CH_2{=}CHCH_2CH_2CH_2CH_3$ **[1]**
(ii) $CH_3CH{=}CHCH_2CH_2CH_3$ **[1]**
(iii) $CH_3CH_2CH{=}CHCH_2CH_3$ **[1]**
(iv) $CH_3CH{=}CCH_2CH_2CH_3$ **[1]**
         |
        CH_3
(v) $CH_2{=}CHCH_2CHCH_2CH_3$ **[1]**
           |
        CH_3

(b) **(i)** but-2-ene **[1]**

(ii) but-1-ene **[1]**

(iii) 2-methylbut-1-ene **[1]**

(iv) pent-2-ene **[1]**

(c) A functional group is always located by the lower number of carbon atom in the chain. **[1]** The double bond joins carbons 2 and 3 in the chain so it is but-2-ene, or joins carbons 1 and 2 (not 3 and 4) so is but-1-ene. **[1]**

(d) **(i)** $CH_2=CHCH_2CH_2CH_3$,

$CH_3CH=CHCH_2CH_3$ **[2]**

(ii) Any two from

$CH_3CH_2C=CH_2$ **[1]**
$\qquad\quad |$
$\qquad\quad CH_3$

$CH_3CH=C-CH_3$ **[1]**
$\qquad\qquad\;\; |$
$\qquad\qquad\;\; CH_3$

$CH_2=CH-CH-CH_3$ **[1]**
$\qquad\qquad\quad |$
$\qquad\qquad\quad CH_3$

15.3

$CH_3-CH_2-CH_2-CH_2-CH_2-CH-CH_3$ 2-methylheptane
$\qquad\qquad\qquad\qquad\qquad\qquad\quad |$
$\qquad\qquad\qquad\qquad\qquad\qquad\quad CH_3$

$CH_3-CH_2-CH_2-CH_2-CH-CH_2-CH_3$ 3-methylheptane
$\qquad\qquad\qquad\qquad\qquad\;\; |$
$\qquad\qquad\qquad\qquad\qquad\;\; CH_3$

$CH_3-CH_2-CH_2-CH-CH_2-CH_2-CH_3$ 4-methylheptane
$\qquad\qquad\qquad\qquad\; |$
$\qquad\qquad\qquad\qquad\; CH_3$

$CH_3-CH_2-CH_2-CH-CH-CH_3$ 2,3-dimethylhexane
$\qquad\qquad\qquad\qquad\; | \quad\; |$
$\qquad\qquad\qquad\qquad\; CH_3\; CH_3$

$CH_3-CH_2-CH-CH_2-CH-CH_3$ 2,4-dimethylhexane
$\qquad\qquad\quad\; | \qquad\qquad |$
$\qquad\qquad\quad\; CH_3\qquad\quad CH_3$

$CH_3-CH_2-CH-CH-CH_2-CH_3$ 3,4-dimethylhexane
$\qquad\qquad\quad\; | \quad\; |$
$\qquad\qquad\quad\; CH_3\; CH_3$

$CH_3-CH-CH_2-CH_2-CH-CH_3$ 2,5-dimethylhexane
$\qquad\; | \qquad\qquad\qquad\; |$
$\qquad\; CH_3 \qquad\qquad\quad CH_3$

$\qquad\qquad\qquad CH_3$
$\qquad\qquad\qquad |$
$CH_3-CH_2-CH_2-C-CH_2-CH_3$ 3,3-dimethylhexane
$\qquad\qquad\qquad |$
$\qquad\qquad\qquad CH_3$

$\qquad\qquad\qquad\qquad CH_3$
$\qquad\qquad\qquad\qquad |$
$CH_3-CH_2-CH_2-CH_2-C-CH_3$ 2,2-dimethylhexane
$\qquad\qquad\qquad\qquad |$
$\qquad\qquad\qquad\qquad CH_3$

$\qquad\qquad\qquad CH_3$
$\qquad\qquad\qquad |$
$CH_3-CH_2-CH-C-CH_3$ 2,2,3-trimethylpentane
$\qquad\qquad\; | \quad |$
$\qquad\qquad\; CH_3\; CH_3$

$\qquad\qquad\qquad\quad CH_3$
$\qquad\qquad\qquad\quad |$
$CH_3-CH-CH_2-C-CH_3$ 2,2,4-trimethylpentane
$\qquad\;\; | \qquad\qquad |$
$\qquad\;\; CH_3 \qquad\quad CH_3$

$\qquad\qquad\quad CH_3$
$\qquad\qquad\quad |$
$CH_3-CH-CH-CH-CH_3$ 2,3,4-trimethylpentane
$\qquad\;\; | \qquad\;\; |$
$\qquad\;\; CH_3 \qquad CH_3$

$\qquad\qquad\quad CH_3$
$\qquad\qquad\quad |$
$CH_3-CH_2-C-CH_2-CH_3$ 2,3,3-trimethylpentane
$\qquad\qquad\;\; | \quad\; |$
$\qquad\qquad\;\; CH_3\; CH_3$

$\qquad\qquad\quad CH_3$
$\qquad\qquad\quad |$
$CH_3-CH_2-C-CH_2-CH_3$ 3-ethyl-3-methylpentane
$\qquad\qquad\;\; |$
$\qquad\qquad\;\; CH_2$
$\qquad\qquad\;\; |$
$\qquad\qquad\;\; CH_3$

$\qquad\; CH_3 CH_3$
$\qquad\;\; | \quad\; |$
$CH_3-C-C-CH_3$ 2,2,3,3-tetramethylbutane
$\qquad\;\; | \quad\; |$
$\qquad\; CH_3 CH_3$

15.4 $CH_3-CH_2-CH_2-CH=CH_2$ pent-1-ene

$CH_3-CH_2-CH=CH-CH_3$ pent-2-ene

$CH_3-C=CH-CH_3$ 2-methylbut-2-ene
$\qquad\;\; |$
$\qquad\;\; CH_3$

$CH_3-CH-CH=CH_2$ 3-methylbut-1-ene
$\qquad\;\; |$
$\qquad\;\; CH_3$

$CH_3-CH_2-C=CH_2$ 2-methylbut-1-ene **[5]**
$\qquad\qquad\; |$
$\qquad\qquad\; CH_3$

15.5 **(a)** **(i)** $CH_3-CH_2-CH_2-CH_3$ **[1]**

(ii) $CH_3-CH_2-CH_3$ **[1]**
$\qquad\qquad\;\; |$
$\qquad\qquad\;\; CH_3$

(b) **(i)** C_4H_{10} **[1]**

(ii) Both have the same molecular formula but different structural formulae. **[1]**

15.6 but-1-ene, but-2-ene, 2-methylpropene. **[3]**

---

16 Nomenclature and formulae of open chain compounds containing functional groups

16.1 **(a)** **(i)** propan-1-ol **[1]**

(ii) propan-2-ol **[1]**

(iii) 2-methylpropan-2-ol **[1]**

(iv) 3-methylpentan-3-ol **[1]**

(v) pentan-2-ol **[1]**

(b) (i) CH_3CH_2OH **[1]**

(ii) $CH_3CH_2CH_2OH$ **[1]**

(iii) CH_3CHCH_3 **[1]**
　　|
　　OH

(iv) $CH_3CH_2CH_2CH_2OH$ **[1]**

(v) $CH_3CH_2CHCH_3$ **[1]**
　　　|
　　　OH

(c) Because it is the same as butan-2-ol. **[1]**

16.2 (a) propan-1,3-diol, 1,3-dihydroxypropane **[2]**

(b) butan-2,3-diol, 2,3-dihydroxybutane **[2]**

(c) butan-1,3-diol, 1,3-dihydroxybutane **[2]**

(d) 2-methylpropan-1,2-diol, 1,2-dihydroxy-2-methylpropane **[2]**

(e) It may be slightly more longwinded to use the prefix, but it can be more straightforward if there are side chains as well. **[1]**

16.3 (a) 1-chlorobutane **[1]**

(b) 2-chlorobutane **[1]**

(c) 1-chloropropane **[1]**

(d) 2,4-dichloropentane **[1]**

(e) 1,2-dichloro-2-methylpropane **[1]**

16.4 (a) (i) 1,4-dibromobutane **[1]**

(ii) 1,3-dibromobutane **[1]**

(iii) 1,1-dibromopropane **[1]**

(iv) 2,4-dibromopentane **[1]**

(v) 1,2-dibromo-2-methylpropane **[1]**

(b) **(i), (ii)** and **(v)**. **[1]**

16.5 (a) $CH_3CH_2CH_2CHO$ **[1]**

(b) $CH_3CCH_2CH_3$ **[1]**
　　　||
　　　O

(c) $CH_3CH_2CH_2CH_2CHO$ **[1]**

(d) $CH_3C\ CH_2CH_2CH_3$ **[1]**
　　　||
　　　O

$CH_3CH_2CCH_2CH_3$ **[1]**
　　　　||
　　　　O

(e) $CH_3CH_2CH_2COOH$ **[1]**

(f) $CH_3CH_2CHCOOH$ **[1]**
　　　　|
　　　　OH

CH_3CHCH_2COOH **[1]**
　　|
　　OH

$HOCH_2CH_2CH_2COOH$ **[1]**

16.6 (a) (i) $CH_3COOCH_2CH_3$ **[1]**

(ii) CH_3COOCH_3 **[1]**

(iii) $HCOOCH_3$ **[1]**

(iv) $HCOOCH_2CH_3$ **[1]**

(b) (ii) and **(iv)** **[1]**

16.7 (a) $CH_3CH_2NH_2$ **[1]**

(b) CH_3CHCH_3 **[1]**
　　|
　　NH_2

(c) CH_3CONH_2 **[1]**

(d) $CH_3CH_2CONH_2$ **[1]**

17 Reactions of alkanes and alkenes and bonding between carbon atoms

17.1 (a) CH_2ClCH_2Cl **[1]**

(b) $CH_3CHClCH_2Cl$ **[1]**

(c) CH_3CH_2Cl

(d) mainly $CH_3CHClCH_3$ and a small amount of $CH_3CH_2CH_2Cl$

(e) $CH_2ClCHClCHClCH_2Cl$

17.2 (a) Initial step is breaking $C{=}C$. **[1]**

First bond formed is between C and H from HCl. **[1]**

$CH_3C^+HCH_3$ is formed as it is a more stable intermediate than $CH_3CH_2C^+H_2$. **[1]**

Because of the inductive effect of carbon it is able to donate more charge than hydrogen as it has more electrons around it. **[1]**

Cl^- then joins to the C^+. **[1]**

(b) The first step is the formation of the more stable $CH_2{=}CHC^+HCH_2Br$ and Br^- the other isomer or ion being $CH^+_2CH{=}CHCH_2Br$. **[1]**

The stable carbonium ion or carbocation is stabilized further by delocalization. **[1]** The Br^- then joins to the C^+ of the tautomer. This spreads the two large groups better. **[1]**

17.3

(a)

$$\begin{matrix} & H & H & \\ & | & | & \\ H- & C & - C & -H \quad \text{all } 109° \\ & | & | & \\ & H & H & \end{matrix}$$ **[1]**

(b)

$$\begin{matrix} H & & H \\ \diagdown & & \diagup \\ & C{=}C & \\ \diagup & & \diagdown \\ H & & H \end{matrix} \quad \begin{matrix}\text{all } 120° \\ \text{(approximately)}\end{matrix}$$ **[1]**

(c) $H-C{\equiv}C-H$ all 180° **[1]**

(d)

$$\begin{matrix} & H & \\ & {}_{109+}|{}_{109-} & \\ Cl & -C- & H \\ & {}_{109+}|{}_{109-} & \\ & H & \end{matrix}$$ **[1]**

(e)

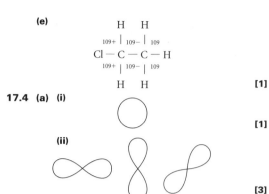

[1]

17.4 (a) (i)

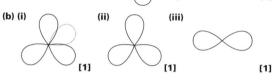

[1]

(ii)

[3]

(b) (i)   **(ii)**   **(iii)**

[1]   **[1]**   **[1]**

(c) (i) sp^3, **(ii)** sp^3, **(iii)** sp^2, **(iv)** sp. **[4]**

17.5 (a) tetrahedral **[1]**

(b) two joined tetrahedra **[1]**

(c) planar **[1]**

(d) tetrahedral **[1]**

(e) four tetrahedra arranged tetrahedrally around the central carbon. **[1]**

17.6 (a) bromomethane,
dibromomethane,
tribromomethane,
tetrabromomethane. **[4]**

(b) free radical substitution **[1]**

(c) (i) $Cl—Cl \rightarrow Cl^{\bullet} + Cl^{\bullet}$ **[1]**

(ii) $Cl^{\bullet} + H—CH_3 \rightarrow HCl + {}^{\bullet}CH_3$ **[1]**
${}^{\bullet}CH_3 + Cl—Cl \rightarrow CH_3Cl + Cl^{\bullet}$ **[1]**

(iii) any two from $Cl^{\bullet} + Cl^{\bullet} \rightarrow Cl—Cl$ **[1]**
${}^{\bullet}CH_3 + {}^{\bullet}CH_3 \rightarrow CH_3—CH_3$ **[1]**
${}^{\bullet}CH_3 + {}^{\bullet}Cl \rightarrow CH_3—Cl$ **[1]**

17.7 (a) $CH_4 + 2O_2 \rightarrow CO_2 + 2H_2O$

(b) $C_3H_8 + 5O_2 \rightarrow 3CO_2 + 4H_2O$

(c) $C_8H_{18} + 12\frac{1}{2}O_2 \rightarrow 8CO_2 + 9H_2O$

(d) $C_3H_6 + 4\frac{1}{2}O_2 \rightarrow 3CO_2 + 3H_2O$

(e) $C_5H_8 + 7O_2 \rightarrow 5CO_2 + 4H_2O$

17.8 (a) appropriate axes **[1]**
plotting **[1]**
line of best fit **[1]**
stating there is a regular or linear relationship
[1]

(b) 4800–4900 kJ mol^{-1}

(c) the most highly branched alkane **[1]**

$$CH_3—\overset{\displaystyle CH_3CH_3}{\underset{\displaystyle CH_3CH_3}{\overset{|\quad|}{\underset{|\quad|}{C—C}}}}—CH_3$$

[1]

17.9 (a)

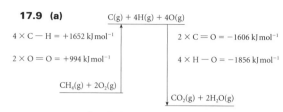

$4 \times C—H = +1652$ kJ mol^{-1}
$2 \times O=O = +994$ kJ mol^{-1}
$2 \times C=O = -1606$ kJ mol^{-1}
$4 \times H—O = -1856$ kJ mol^{-1}

total endothermic $= +2646$ kJ mol^{-1} **[1]**
total exothermic $= -3462$ kJ mol^{-1} **[1]**
overall exothermic $= -816$ kJ mol^{-1} **[1]**

(b) (i) LHS correctly drawn and balanced **[1]**
RHS correctly drawn and balanced **[1]**
endothermic
$8 \times C—H = 3304$ kJ mol^{-1}
$2 \times C—C = 692$ kJ mol^{-1}
$5 \times O=O = 2485$ kJ mol^{-1}
total $= 6481$ kJ mol^{-1} **[1]**
exothermic
$6 \times C=O = -4818$ kJ mol^{-1}
$8 \times H—O = -3712$ kJ mol^{-1}
total $= -8530$ kJ mol^{-1} **[1]**
overall it is exothermic
$= -2049$ kJ mol^{-1} **[1]**
It is 170 kJ mol^{-1} different and less exothermic. **[1]**

(ii) LHS correctly drawn and balanced **[1]**
RHS correctly drawn and balanced **[1]**
endothermic
$18 \times C—H = 7434$ kJ mol^{-1}
$7 \times C—C = 2422$ kJ mol^{-1}
$12\frac{1}{2} \times O=O = 6213$ kJ mol^{-1}
total $= 16089$ kJ mol^{-1} **[1]**
exothermic
$16 \times C=O = -12848$ kJ mol^{-1}
$18 \times H—O = -8352$ kJ mol^{-1}
total $= 21200$ **[1]**
overall it is exothermic
$= -5111$ **[1]**
It is 401 kJ mol^{-1} different and less exothermic. **[1]**

18 Reactions of alcohols

18.1 (a) chloroethane or ethyl chloride **[1]**

(b) 1-chloropropane **[1]**

(c) 2-chloropropane **[1]**

(d) 1-chlorohexane, 2-chlorohexane, 3-chlorohexane **[3]**

(e) 1-chloropentane, 2-chloropentane, 3-chloropentane,
1-chloro-3-methylbutane,
2-chloro-3-methylbutane,

1-chloro-2-methylbutane,
2-chloro-2-methylbutane,
1-chloro-2,2-dimethylpropane **[8]**

18.2 (a) (i) $CH_3CH_2CH_2I + H_2O$ **[1]**

(ii) $CH_3CH_2CH_2Br + H_2O$ **[1]**

(iii) $CH_3 \, CH \, CH_3 + H_2O$ **[1]**
  $|$
  Br

(b) Concentrated sulphuric acid 'absorbs' the water which is produced. **[1]**
This pulls the reaction to the right or stops it returning back to the left. **[1]**

18.3 (a) $CH_3CH{=}CH_2$ **[1]**

(b) $CH_3CH{=}CHCH_3 + CH_2{=}CHCH_2CH_3$ **[2]**

(c) $CH_3CH_2CH{=}CH_2$ **[1]**

18.4 (a) Cl^-, Br^-, I^- **[1]**

(b) (i) $Br^- \, C{-}OH \rightarrow Br{-}C + OH^-$

[1 mark to show attack and elimination at the same time.] [1 mark to show attack from opposite side to OH⁻ leaving.]

(ii) $C{-}OH \rightarrow C^+ + OH^-$
  $C^+ + Br^- \rightarrow CBr$

[1 mark to show bond breaking first.] [1 mark to show attack from either side of carbonium ion.]

(iii) Relieves steric strain going from 109° to 120°. Easier to attack with only H's in the way. **[2]**

(iv) S_N1 can also undergo elimination. **[1]**

18.5 (a)

CH_3OH	methanoic acid	HCOOH	**[3]**
CH_3CH_2OH	ethanoic acid	CH_3COOH	**[3]**
$CH_3CH_2CH_2OH$	propanoic acid	CH_3CH_2COOH	**[3]**
CH_3CHCH_3 $\quad|$ $\quad OH$	propanone	CH_3CCH_3 $\quad\|\|$ $\quad O$	**[3]**
CH_3 $\quad|$ CH_3CCH_3 $\quad|$ $\quad OH$	no reaction		**[2]**

(b)

CH_3OH	methanal	HCHO	**[2]**
CH_3CH_2OH	ethanal	CH_3CHO	**[2]**
$CH_3CH_2CH_2OH$	propanal	CH_3CH_2CHO	**[2]**
CH_3CHCH_3 $\quad|$ $\quad OH$	propanone	CH_3CCH_3 $\quad\|\|$ $\quad O$	**[2]**
CH_3 $\quad|$ CH_3CHCH_3 $\quad|$ $\quad OH$	no reaction		

19 Reactions of halogenoalkanes and halogenoalkenes

19.1 (a) Indicating any **four** from:
first involves breaking C—Br bond,
leaves C^+,
the ion is planar,
attack by OH^- can be from either side,
the product will be racemic or any optical activity is lost. **[4]**

(b) Indicating any **four** from:
attack is by OH^-,
attack is from opposite side of molecule to C—Br bond,
new bond is forming as old bond is breaking,
molecule is inverted or stereochemistry is inverted,
optical activity is retained. **[4]**

19.2 (a) (i) $CH_3CH_2CH_2OH$ and $CH_3CH{=}CH_2$ **[2]**

(ii) CH_3CHCH_3 and $CH_3CH{=}CH_2$ **[2]**
  $\quad|$
  $\quad OH$

(iii) $CH_3CHCH_2CH_3$,
  $\quad|$
  $\quad OH$
  $CH_3CH{=}CH_2CH_3$ and
  $CH_2{=}CHCH_2CH_3$ **[3]**

(b) Any two from:
weak bases,
lower temperatures,
water rather than an organic solvent, if NaOH used. **[2]**

(c) Elimination is the favoured as steric strain is relieved in the product. The groups become less crushed together. **[1]**

19.3 (a) chloroethene **[1]**

(b) (i) addition **[1]**

(ii) elimination **[1]**

(c) ethyne **[1]**

(d) ethane-1,2-diol and 2-chloroethanol. **[2]**

19.4 (a) butan-1-ol and but-1-ene **[2]**

(b) cream from bromobutane
white from chlorobutane
yellow from iodobutane **[3]**

(c) Iodobutane will hydrolyse the fastest **[1]** as the C—I bond is the weakest or longest **[1]** and so most easy to break. **[1]**

(d) (i) 1.37 g **[1]**

(ii) 1.03 g **[1]**

Part 2 A2 questions

20 Energetics 2

20.1 The lattice enthalpy of an ionic crystal is the standard enthalpy change of formation of one mole of the crystalline lattice from its constituent ions in the gas phase. **[3]**

20.2 (a) The first ionization energy of an element, E_{m1}, is the energy required to remove one mole of electrons from one mole of atoms of the element in the gaseous state to form singly-charged gaseous ions. **[3]**

e.g. $M(g) \rightarrow M^+(g) + e^-$, $E_{m1} = +v\,kJ\,mol^{-1}$ **[1]**

(b) The second ionization energy of an element, E_{m2}, is the energy required to remove one mole of electrons from one mole of singly-charged ions in the gaseous state to form doubly-charged gaseous ions. **[3]**

e.g. $M^+(g) \rightarrow M^{2+}(g) + e^-$, $E_{m2} = +w\,kJ\,mol^{-1}$ **[1]**

(c) The first electron affinity of an element, E_{aff1}, is the energy change occurring when one mole of atoms in the gaseous state accepts one mole of electrons to form one mole of singly-charged gaseous ions. **[3]**

e.g. $R(g) + e^- \rightarrow R^-(g)$; $E_{aff1} = -x\,kJ\,mol^{-1}$ **[1]**

(d) The standard enthalpy change of atomization of an element, ΔH_{at}, is the enthalpy change that takes place when one mole of gaseous atoms is made from the element in its standard physical state under standard conditions. **[3]**

e.g. $M(s) \rightarrow M(g)$; $\Delta H_{at}^{\ominus} = +y\,kJ\,mol^{-1}$

or

$\frac{1}{2}R_2(g) \rightarrow R(g)$; $\Delta H_{at}^{\ominus} = +z\,kJ\,mol^{-1}$ **[1]**

20.3 In the case of the first electron affinity an electron is being attracted by the positively charged nucleus and helps to bring the electron configuration nearer to the stable arrangement of a noble gas. **[1]** However, in the case of the second electron added, even though it is completing the noble-gas configuration for the ion, **[1]** it is being added to an ion which already has a negative charge and so a force of repulsion has to be overcome which requires an energy input. **[1]**

20.4 (a) Hess's law states that the total enthalpy change accompanying a chemical change is independent of the route by which the chemical change takes place. **[3]**

(b) lattice enthalpy $= \Delta H_f^{\ominus}\,[MN(s)] - \Delta H_1$ **[1]**

(c) Standard enthalpy change of atomization of M, standard enthalpy change of atomization of N, first ionization energy of M and the (first) electron affinity of N. **[4]**

20.5 (a) ΔH_1 – enthalpy change of atomization of potassium, ΔH_2 – enthalpy change of atomization of bromine, ΔH_3 – first ionization energy of potassium, ΔH_4 – electron affinity of bromine, ΔH_5 – enthalpy change of formation of solid potassium bromide, ΔH_6 – lattice enthalpy of potassium bromide. **[5]**

(b) $\Delta H_6 = -689.3\,kJ\,mol^{-1}$ **[2]**

20.6 (a) $-793\,kJ\,mol^{-1}$, **(b)** $-2527\,kJ\,mol^{-1}$, **(c)** $-2601\,kJ\,mol^{-1}$, **(d)** $-2950\,kJ\,mol^{-1}$, **(e)** $-15327\,kJ\,mol^{-1}$. **[4 marks for each part.]**

20.7 (a) (i) the lattice energy becomes less negative **[1]**

(ii) the lattice energy becomes less negative **[1]**

(iii) the lattice energy becomes more negative **[1]**

(iv) the lattice energy becomes more negative **[1]**

(b) As the size of the ions increases, the ionic change is less concentrated on the ion, so the force of attraction between the ions diminishes **[1]**: the ionic bonds therefore become weaker **[1]** and, therefore, less energy is given out during their formation. **[1]** As the charge on the ions increases, so does the force of attraction between the ions **[1]**: the ionic bonds therefore become stronger **[1]** and, therefore, more energy is given out during their formation. **[1]**

20.8 (a) CaI $-69.4\,kJ\,mol^{-1}$ **[4]**
CaI_2 $-538.0\,kJ\,mol^{-1}$ **[4]**
CaI_3 $+783.4\,kJ\,mol^{-1}$ **[4]**

(b) CaI_3 is, clearly, energetically unstable both with respect to its elements and to the other iodides,

CaI is energetically stable with respect to its elements but energetically unstable with respect to CaI_2 ($\Delta H^{\ominus}$ for

$$2CaI(s) \rightarrow CaI_2(s) + Ca(s)$$
$$= -390.2 \text{ kJ mol}^{-1}).$$

CaI_2 is energetically stable both with respect to its elements and to the other iodides. **[6]**

20.9 (a) $CuBr -107.4 \text{ kJ mol}^{-1}$ **[4]**

$CuI -70.3 \text{ kJ mol}^{-1}$ **[4]**

$CuBr_2 -146.1 \text{ kJ mol}^{-1}$ **[4]**

$CuI -6.9 \text{ kJ mol}^{-1}$ **[4]**

(b) (i) Comparing the energetic stabilities of CuI and CuI_2:

For

$$4CuI_2(s) \rightarrow 2CuI(s) + 2Cu(s) + 3I_2(s);$$
$$\Delta H^{\ominus} = -113.0 \text{ kJ mol}^{-1}$$

Therefore copper(II) iodide is unstable with respect to decomposition into copper(I) iodide, copper and iodine. Thus copper(II) iodide is not formed when solutions of copper(II) ions and iodide ions are mixed, a mixed precipitate of copper(I) iodide and copper being formed instead. **[4]**

(ii) In **(ii)**, do a similar calculation for the bromides as has been done for the iodides in **(b)(i)**.

For $4CuBr_2(s) \rightarrow 2CuBr(s) + Cu(s) + 3Br_2(l); \Delta H^{\ominus} = +369.6 \text{ kJ mol}^{-1}$

Therefore copper(II) bromide is energetically stable with respect to decomposition. So one would expect that copper(II) bromide would be formed when solutions of copper(II) ions and bromide ions are mixed (assuming, of course, that there is no kinetic barrier to reaction). **[4]**

20.10 (a) LiF 0.6%, LiI 3.7%, NaF 1.4%, NaI 3.2%, KF 1.6%, KI 2.7%, AgF 1.5%, AgI 10.0%, BeF_2 11.3%, BeI_2 5.5%, MgF_2 1.5%, MgI_2 19.7%, CaF_2 0.8%, CaI_2 8.9%, ZnF_2 3.5%, ZnI_2 2.2%. **[4]**

(b) A model based on spherical ions would appear to be highly satisfactory for LiF, NaF, KF, AgF, MgF_2 and CaF_2 (discrepancy between values in all of these cases of 2% or less). **[2]**

(c) A model based on spherical ions would appear to be least satisfactory for AgI, BeF_2, BeI_2, MgI_2 and CaI_2 (a discrepancy of more than 5% in each case).

Ion polarization distorts ions from a spherical shape. It is most significant when the positive ion is small and/or highly charged e.g. Be^{2+} and the negative ion large e.g. I^- and/or highly charged. All those listed in the over 5% discrepancy category contain Be^{2+}, I^- or both. **[5]**

(One can explore the figures more closely still but beware! There are inconsistencies in the picture.)

21 Kinetics

21.1 In this question, the method (or methods) given may not be the only feasible method(s) for following the rate of the particular reaction considered in each part and marking should be carried out bearing this in mind.

(a) Measure the volume of gas produced with time, keeping the temperature constant. **[2]**

(b) Dilatometry – make use of the very small change of volume that takes place in the reaction mixture with time, keeping the temperature constant. **[3]**

(c) Measure the change in pressure of the mixture with time, keeping the temperature constant. **[2]**

(d) Polarimetry – take readings of the extent of rotation of polarized light with time, keeping the temperature constant. **[2]**

(e) Withdraw samples of reaction mixture, quench by cooling and titrate with alkali of known concentration: a 'final' titration is necessary as is the need to make allowance for the amount of alkali that neutralizes the acid catalyst present. **[5]**

(f) Measure the volume of gas produced with time, keeping the temperature constant. **[2]**

(g) Withdraw samples of reaction mixture, quench by neutralizing the acid with sodium hydrogencarbonate, add excess potassium iodide solution and titrate the liberated iodine with sodium thiosulphate solution of known concentration: a 'final' titration is necessary. **[5]**

or

Use a colorimetric method. Mix the reactants together and start a stopclock. Take readings of the colorimeter at known times. Prepare a calibration curve. Use it to obtain values for the concentration of bromine at the measured times. **[5]**

21.2 (a) (i) The rate constant **[1]**: a constant of proportionality linking the rate of reaction with the concentration terms appearing in the rate equation. **[1]**

(ii) The concentrations of substances A and B, respectively **[1]** measured in mol dm^{-3}. **[1]**

(iii) The orders of the reaction with respect to substances A and B; they are the experimentally determined powers to which the concentrations of A and B are raised in the rate equation. **[1]**

(b) Because most chemical reactions take place in a series of steps whereas the stoicheiometric equation deals only with the initial reactants and final products. **[2]**

(c) Consider one reactant at a time. **[1]** Keep the concentrations of all other reactants much larger than the one being considered. **[1]** Vary the concentration of the substance under consideration and study the effect on the rate of reaction. **[1]** Repeat with the other reactants in turn. **[1]**

21.3 (a) Most reactions take place in a series of steps rather than being simple one-step reactions. **[1]** These steps are known as the mechanism of the reaction. **[1]**

(b) The 'rate-determining step'. **[1]**

(c) Mechanisms are not deduced from rate equations. Rate equations must be consistent with a suggested mechanism, however, if the mechanism is to be viable. **[2]**

21.4 Doubling [P], doubles rate, therefore first order with respect to P. **[1]**
Doubling [Q] doubles rate, therefore first order with respect to Q. **[1]**
Therefore consistent with the suggested rate equation. **[1]**

21.5 (a) Comparing the first two sets of data: doubling [X] doubles rate, therefore first order with respect to X. **[1]**
Comparing the second and third or fourth and fifth sets of data: increasing [Y] has no

effect on rate, therefore zero order with respect to Y. **[1]**
Therefore rate equation is: Rate = $k[X] [Y]^0$ or Rate = $k[X]$ (since $[z]^0 = 1$, for any value of z). **[1]** Overall order = first (sum of the separate orders = 1). **[1]**

(b) Value of rate constant = 0.3 **[1]** Substituting units in the rate equation:
$[mol \ dm^{-3} \ s^{-1}] = [units \ of \ k] [mol \ dm^{-3}]$ **[1]**
from which the units of k (the rate constant) = s^{-1} **[1]**

21.6 (a) First, **[1] (b)** Third, **[1] (c)** First, **[1] (d)** Fourth. **[1]**

21.7 Plot graph of concentration against time and show that the half-life for the reaction is constant. The reaction is therefore first order and the suggestion is supported. **[5]**

21.8 (a) Iodine is formed in the reaction and reacts with the starch present in the mixture to give a blue colour. **[2]**

(b) The appearance of the blue colour is a measure of the formation of a fixed amount of iodine. **[1]**

(c) It is the total volume of the mixture that has to be kept constant. **[1]** To ensure that this is so, different volumes of water have to be added. **[1]**

(d) Only the concentration of iodide ions is varied. No information is available on the effect of varying the concentration of hydrogen peroxide or of hydrogen ions. **[1]**

(e) First, vary the concentration of hydrogen peroxide, keeping the concentrations of hydrogen ions and iodide ions constant. Measure the time taken in each run for the appearance of a blue colour. **[2]**
Then, vary the concentration of hydrogen ions, keeping the concentration of iodide ions and hydrogen peroxide constant. Again measure the time taken for the appearance of the blue colour. **[2]**

(f) Rate is proportional to $1/time$. Values of $1/t$ for the four reactions are, respectively: 0.0130, 0.0208, 0.0322 and 0.0455 (all in s^{-1}). **[2]**

(g) A graph of rate against volume of KI added (a measure of the concentration of iodide ions added) is within the limits of experimental error, a straight line indicating a first order reaction. (Don't forget that the graph must go through the origin.) **[4]**

21.9 Rate-determining step shows HBr and HBrO$_3$ involved. **[1]**

But HBr is formed from H$^+$ and Br$^-$ in a previous fast step. **[1]**

Similarly HBrO$_3$ is formed from H$^+$ and BrO$_3^-$ in a previous fast step. **[1]**

Therefore [H$^+$] (twice – and hence a squared term), [Br$^-$] and [BrO$_3^-$] (each once) should all appear in the rate equation. **[1]**

They do so, therefore the rate equation is consistent with the mechanism. **[1]**

21.10 (a) Mix together, at zero time, solutions of propanone, dilute sulphuric acid and iodine. At known times, withdraw samples, quench using sodium hydrogencarbonate solution. Titrate remaining iodine with sodium thiosulphate of known concentration. Plot a graph of titre against time and determine the order of the reaction from the shape of the graph. A straight line indicates zero order, a curve with constant half-life indicates first order and a deeper curve with a considerably increasing half-life indicates second order. **[5]**

and

Make a series of mixtures of hydrochloric acid, propanone and water. Start the reaction by adding the appropriate amount of iodine and measure the time taken for the iodine colour to disappear. In each run, the amount of one substance (hydrochloric acid, propanone or iodine) is varied and the amount of water is adjusted to ensure that the total volume (including that of iodine) remains the same. Examine how changing the amount of a particular substance affects the time and, hence, the rate of the reaction: thus determine the order of reaction with respect to each reactant in turn. **[5]**
Or other suitable method. **[5]**

(b) Take first and third runs: concentration of propanone is halved, rate is halved, therefore first order with respect to propanone, **[1]**
Take first and second runs: concentration of iodine is halved, rate is unaffected, therefore zero order with respect to iodine, **[1]**
Take third and fourth runs: concentration of hydrogen ions is halved, rate is halved, therefore first order with respect to hydrogen ions. **[1]**

(c) Rate = k [propanone] [hydrogen ions] [iodine]^0
or
Rate = k [propanone] [hydrogen ions] **[1]**

(d) A possible mechanism would be:
H$^+$ + CH$_3$COCH$_3$ → CH$_3$COH$^+$CH$_3$ slow, therefore rate-determining
CH$_3$COH$^+$CH$_3$ → CH$_2$=COHCH$_3$ + H$^+$ fast
CH$_2$=COHCH$_3$ + I$_2$ → CH$_2$ICOH(I)CH$_3$ fast
CH$_2$ICOH(I)CH$_3$ → CH$_2$ICOCH$_3$ + H$^+$I$^-$ fast **[5]**

21.11 The rate-determining step of a S$_N$1 reaction involves only the substance R—X. **[1]** The reaction is therefore first order overall. **[1]**
The rate-determining step of a S$_N$2 reaction involves both R—X and the nucleophile (the example used in the question is OH$^-$). **[1]** The reaction is therefore second order overall. **[1]**

21.12 (a) In each case, an accurate graph of titre against time is drawn with axes correctly labelled. The graph for 1-chloro-1-phenylethane should show first order kinetics (constant half-life) and the graph for 1-chloro-2-phenylethane should show second order kinetics (half-lives increase considerably). **[8]**

(b) Overall orders – since neither reactant is present in a large excess. **[2]**

(c) 1-chloro-1-phenylethane – first order – therefore S$_N$1, **[1]**
1-chloro-2-phenylethane – second order – therefore S$_N$2. **[1]**

(d) Rate constant is found by measuring the average half-life from the graph of titre against time and then using the relationship $t_{1/2}$ = ln 2/k. The half-life is approximately 10.0 hours so $t_{1/2}$ = 0.069 s^{-1}.
Alternatively, the gradient can be found at a certain time – this is a measure of the rate of reaction which can then be substituted into the rate equation,
Rate = k[1-chloro-1-phenylethane]
using the appropriate concentration for 1-chloro-1-phenylethane which is equal to [OH$^-$].
Another possible method is to plot in [1-chloro-1-phenylethane] against time which will give a straight line graph of gradient $-k$. **[4]**

22 Equilibrium 2

22.1 (a) The equilibrium law is an empirical law, that is, a very large number of experimental investigations have shown that 'for any system at equilibrium, there is a simple relationship between the concentrations of the substances present'. **[2]**

(b) For the reaction shown,

$$K_c = \frac{[C]^p[D]^q}{[A]^m[B]^n}$$

where K_c is known as the equilibrium constant. **[2]** It has a constant value at a given temperature.

(c) When K_c is very large, the reaction goes virtually to completion. **[1]**
When K_c is very small, the reaction barely takes place. **[1]**

22.2 (a) $\dfrac{[C_2H_5CO_2CH_3(l)]\,[H_2O(l)]}{[CH_3OH(l)]\,[C_2H_5OH(l)]}$ (no units) **[3]**

(b) $\dfrac{[CH_3CO_2C_5H_{11}(l)]}{[C_5H_{10}(l)]\,[CH_3CO_2H(l)]}$ (dm³ mol⁻¹) **[3]**

(c) $\dfrac{[O_2(g)]^3}{[O_3(g)]^2}$ (mol dm⁻³) **[3]**

(d) $\dfrac{[NO_2(\text{trichloroethane})]^2}{[N_2O_4(\text{trichloroethane})]}$ (mol dm⁻³) **[3]**

(e) $\dfrac{[Cu(NH_3)_4^{2+}(aq)]}{[Cu^{2+}(aq)]\,[NH_3(aq)]^4}$ (dm¹² mol⁻⁴) **[3]**

22.3 (a) (i) $\dfrac{[p_{PCl_3}][p_{Cl_2}]}{[p_{PCl_5}]}$ (atm) **[3]**

(ii) $\dfrac{[p_{COBr_2}]}{[p_{CO}]\,[p_{Br_2}]}$ (atm⁻¹) **[3]**

(iii) $\dfrac{[p_{P_4}][p_{F_2}]^{10}}{[p_{PF_5}]^4}$ (atm⁷) **[3]**

(iv) $\dfrac{[p_{H_2}]^2\,[p_{S_2}]}{[p_{H_2S}]^2}$ (atm) **[3]**

(v) $\dfrac{[p_{NO}]^4\,[p_{H_2O}]^6}{[p_{NH_3}]^4\,[p_{O_2}]^5}$ (atm) **[3]**

(b) Equilibrium systems in which all the substances involved are in the same phase i.e. all are gases, all are in aqueous solution, etc. **[1]**

22.4 (a) Equilibrium systems in which the substances involved are not all in the same phase. **[1]**

(b) (i) $\dfrac{1}{[Br_2(g)]}$ **[2]**

(ii) $\dfrac{[Zn^{2+}(aq)]}{[Cu^{2+}(aq)]}$ **[2]**

(iii) $[Ag^+(aq)]^3\,[PO_4^{3-}(aq)]$ **[2]**

(iv) $\dfrac{[Fe^{3+}(aq)]}{[Ag^+(aq)]\,[Fe^{2+}(aq)]}$ **[2]**

(c) (i) p_{CO_2} **[2]**

(ii) $[p_{NH_3}]^2[p_{CO_2}]$ **[2]**

22.5 (a) 0.414 **[1]**

(b) Because the volume terms all cancel out when substituted into the equilibrium equation. **[1]**

(c) $K_c = \dfrac{[0.414\ \text{mol/V dm}^3]\,[0.414\ \text{mol/V dm}^3]}{[0.586\ \text{mol/V dm}^3]\,[0.086\ \text{mol/V dm}^3]}$

= 3.4 at 400 K **[3]**

(d) (i) It provides information about the position of equilibrium. A value of 3.4 for the equilibrium constant of this reaction tells us that the position of equilibrium lies near the middle but with a slight excess of products (the substances on the right of the chemical equation) over reactants. **[2]**

(ii) The reaction does not take place under those conditions. **[1]**

(iii) The reaction goes to completion. **[1]**

22.6 Amount of sulphur dichloride dioxide initially = 2.70 g/135 g mol⁻¹ = 0.020 mol in 500 cm³ = 0.040 mol dm⁻³.
Amount of sulphur dichloride dioxide at equilibrium = 2.16 g/135 g mol⁻¹ = 0.016 mol in 500 cm³ = 0.032 mol dm⁻³.
Therefore, no. of moles of sulphur dioxide and chlorine at equilibrium = (0.040 − 0.032) = 0.008 mol dm⁻³.

Therefore, at 300 K,

$$K_c = \frac{[0.008\ \text{mol dm}^{-3}][0.008\ \text{mol dm}^{-3}]}{[0.032\ \text{mol dm}^{-3}]}$$

= 2.00 × 10⁻³ mol dm⁻³ at 300 K. **[5]**

22.7 (a) The partial pressure of each constituent gas in a mixture of gases is the pressure which that gas would exert if it alone occupied the same container. **[2]**

(b) $K_p = \dfrac{[p_{SO_3}]}{[p_{SO_2}]^2[p_{O_2}]}$ **[2]**

(c) $K_p = \dfrac{[0.364 \text{ atm}]^2}{[0.456 \text{ atm}]^2[0.180 \text{ atm}]}$

$= 3.54 \text{ atm}^{-1}$ at 1000 K. **[3]**

(d) (i) Partial pressure of sulphur trioxide will increase,
partial pressure of sulphur dioxide will decrease. **[1]**
K_p will be unchanged in value. **[1]**

(ii) Partial pressures of sulphur trioxide and oxygen will increase,
partial pressure of sulphur trioxide will decrease. **[1]**
K_p will be unchanged in value. **[1]**

(iii) Partial pressure of sulphur trioxide will increase,
partial pressures of sulphur dioxide and oxygen will decrease. **[1]**
K_p will increase in value. **[1]**

(iv) Partial pressures of sulphur dioxide and oxygen will increase,
partial pressure of sulphur trioxide will decrease. **[1]**
K_p will be unchanged in value. **[1]**

22.8 (a) $2\text{Na}(g) \rightleftharpoons \text{Na}_2(g)$ or $\text{Na}_2(g) \rightleftharpoons 2\text{Na}(g)$ **[2]**

(b) $K_p = \dfrac{p_{\text{Na}_2}}{[p_{\text{Na}}]^2}$ or $K_p' = \dfrac{[p_{\text{Na}}]^2}{p_{\text{Na}_2}}$ **[2]**

(c) 71.3 grams of Na atoms $= 3.1$ moles of Na atoms,
28.7 grams of Na_2 molecules $= 0.62$ moles of Na_2 molecules,
therefore partial pressure of Na $= 8.33$ atm and partial pressure of $\text{Na}_2 = 1.67$ atm:
$K_p = 0.024 \text{ atm}^{-1}$ at 1000 K or K_p'
$= 41.6$ atm at 1000 K. **[3]**

22.9 (a) (i) Move towards the right as there are fewer gaseous molecules on the right-hand side of the equation. **[2]**

(ii) Move towards the right as there are fewer gaseous molecules on the right-hand side of the equation. **[2]**

(iii) Unaffected as there are the same number of gaseous molecules on both sides of the equation. **[2]**

(iv) Move towards the right as there are more gaseous molecules on the right-hand side of the equation. **[2]**

(v) Unaffected as there are the same number of gaseous molecules on both sides of the equation. **[2]**

In none of the cases, is there a change in the value of K_p. When there is a change in the number of molecules during the reaction, the equilibrium position will shift as a result of the change in pressure, but the new equilibrium established will be such that the value of the equilibrium constant remains the same as before. **[2]**

(b) (i) exothermic reaction in forward direction, increased temperature favours reaction to the left – K_c decreases, **[3]**

(ii) endothermic reaction in forward direction, reduced temperature favours reaction to the left – K_c decreases, **[3]**

(iii) as (ii), **[3]**

(iv) endothermic reaction in forward direction, increased temperature favours reaction to the right – K_c increases, **[3]**

(v) exothermic reaction in forward direction, decreased temperature favours reaction to the right – K_c increases. **[3]**

22.10 (i) K_p increases with increase in temperature – forward reaction is endothermic, **[2]**

(ii) K_p decreases with increase in temperature – forward reaction is exothermic, **[2]**

(iii) lg K_p, and therefore K_p, increases with increase in temperature – forward reaction is endothermic, **[2]**

(iv) K_a decreases with increase in temperature – forward reaction is exothermic. **[2]**

22.11 (a) A strong acid is one which is almost completely ionized in a solution of moderate dilution. **[2]**

(b) A weak acid is one which is ionized only slightly (and therefore exists mostly in covalently bonded form) in a solution of moderate dilution. **[2]**

(c) A strong alkali is one which exists almost completely as ions in a solution of moderate dilution. **[2]**

(d) A weak alkali is one which is ionized only slightly (and therefore exists mostly in covalently bonded form) in a solution of moderate dilution. **[2]**

22.12 (a) (i) 2 (ii) 1.7 (iii) 3 (iv) 2 (v) 2.3
[2 marks each.]

(b) (i) 12 (ii) 12.3 (iii) 11 (iv) 11.7
(v) 10.8 **[2 marks each.]**

22.13 (a) -0.30 **(b)** -0.30 **(c)** 14.6 **(d)** 14.4
[3 marks each.]

22.14 (a) (i) 0.10 mol dm^{-3} **(ii)** $0.001 \text{ mol dm}^{-3}$
(iii) $0.013 \text{ mol dm}^{-3}$ **(iv)** $1.0 \times 10^{-12} \text{ mol dm}^{-3}$
(v) $4.6 \times 10^{-10} \text{ mol dm}^{-3}$ **[1 mark each.]**

(b) for **(a)(iv)**, hydroxide ion concentration is
0.01 mol dm^{-3}, for **(a)(v)**, hydroxide ion
concentration is $2.17 \times 10^{-5} \text{ mol dm}^{-3}$ **[2]**

22.15 (a) $CH_3CHOHCO_2H(aq) \rightleftharpoons$
$CH_3CHOHCO_2^-(aq) + H^+(aq)$

$$K_a = \frac{[CH_3CHOHCO_2^-(aq)][H^+(aq)]}{[CH_3CHOHCO_2H(aq)]}$$

Therefore $K_a = 1.4 \times 10^{-4} \text{ mol dm}^{-3}$

$$= \frac{[H^+(aq)]^2}{[0.001]}$$

(assume $[H^+(aq)]$ is small)

from which
$[H^+(aq)] = 3.74 \times 10^{-4} \text{ mol dm}^{-3}$
and pH $= 3.4$ **[4]**

(b) pH $= 1.9$ **[3]**

(c) pH $= 2.9$ **[3]**

(d) pH $= 3.9$ **[3]**

22.16 (a) pH $= 3.37$, therefore
$[H^+(aq)] = 4.27 \times 10^{-4} \text{ mol dm}^{-3}$,
therefore $K_a = 1.82 \times 10^{-5} \text{ mol dm}^{-3}$ **[4]**

(b) $K_a = 6.59 \times 10^{-5} \text{ mol dm}^{-3}$ **[3]**

(c) $K_a = 6.31 \times 10^{-10} \text{ mol dm}^{-3}$ **[3]**

(d) $K_a = 1.45 \times 10^{-5} \text{ mol dm}^{-3}$ **[3]**

22.17 (a) $CH_3CO_2^-Na^+(aq) + HCl(aq) \rightarrow$
$CH_3CO_2H(aq) + H_2O(l)$ **[1]**

(b) buffer solution **[1]**

(c) pH $= pK_a - \lg\dfrac{[\text{acid}]}{[\text{base}]}$

therefore for the first addition,

$5.0 = pK_a - \lg\dfrac{[0.02]}{[0.10]}$

therefore $pK_a = 4.70$ therefore
$K_a = 2.00 \times 10^{-5} \text{ mol dm}^{-3}$ **[4]**
for the second addition, K_a
$= 1.89 \times 10^{-5} \text{ mol dm}^{-3}$,
for the third addition, K_a
$= 2.10 \times 10^{-5} \text{ mol dm}^{-3}$,
and for the fourth addition, K_a
$= 2.00 \times 10^{-5} \text{ mol dm}^{-3}$. **[3]**

(d) The method depends on judgement in
matching closely similar colours which is
unlikely to be very accurate. Moreover since
pH is a log scale, any error in its value will be
magnified when converting from a log scale
to a normal scale. **[2]**

(e) It would be best to quote an average value for
K_a. **[1]**

22.18 (a) 5.37 **[2] (b)** 9.25 **[2] (c)** 3.50 **[2]**

(d) 7.68 **[2] (e)** 2.86 **[2]**

(f) (i) 2.30 **[2]**

(ii) The equilibria present are:
$HF(aq) \rightleftharpoons H^+(aq) + F^-(aq)$ (X)
and: $H_2O(l) \rightleftharpoons H^+(aq) + OH^-(aq)$ (Y)

(A) Adding acid will cause (X) to move
to the left resulting in the removal of
the added $H^+(aq)$ ions and so the
pH will change but little.

(B) Adding alkali will remove $H^+(aq)$
ions from (X) by combining with
the added $OH^-(aq)$ i.e. shifting
equilibrium (Y) to the left. More HF
will then ionize until effectively all
the hydroxide ions have been
removed and the pH will therefore
not change significantly. **[5]**

23 Periodic trends in reactivity

23.1 (a) aluminium $<$ magnesium $<$ sodium **[1]**

(b) Reaction involves formation of the metal ion
[1] this becomes harder Na to Al **[1]** as the
nuclear charge increases. **[1]**

23.2 (a) Heated **[1]** in steam. **[1]**

(b) $Mg(s) + H_2O(g) \rightarrow MgO(s) + H_2(g)$ **[2]**

23.3 (a) Al_2Cl_6 ($AlCl_3$), $SiCl_4$, PCl_3, PCl_5 **[4]**

(b) The phosphorus atom has 10 electrons in its
valence shell. **[1]** These cannot be
accommodated in the s and p orbitals
(max. 8) **[1]** so d-orbitals must be involved
and first occur in the 3rd shell (i.e. in P and
not in Al or Si). **[1]**

23.4 (a) Na_2O **[2]**

(b) (i) Na_2O_2 **[2]**

(ii) O_2^{2-} **[1]** peroxide. **[1]**

(c) Lithium is a smaller cation than sodium **[1]**
and the peroxide ion is relatively large. **[1]**
The mismatch in size between the lithium
and peroxide ions may make the lattice
energy of lithium peroxide rather low **[1]**
favouring the simple oxide. (Also allow
argument based on polarization of the large
peroxide ion by the small lithium cation.)

23.5 (a) $PCl_5(s) + 4H_2O(l) \rightarrow H_3PO_4(aq) +$
$5HCl(aq)$ **[2]**

(b) The acid contains three moles of ionizable hydrogen per mole of acid/has three acidic hydrogens. **[1]** Only hydrogens bonded to oxygen (not directly to phosphorus) are ionizable **[1]** so the structure is $O=P(OH)_3$. **[1]**

23.6 Phosphorus(III) chloride is a covalent liquid **[1]** and hence has no ions **[1]** or mobile electrons **[1]** to carry a current. **[1]** It is hydrolysed by water **[1]**:
$PCl_3(l) + 3H_2O(l) \rightarrow H_3PO_3(aq) + 3HCl(aq)$ **[2]** to give a solution containing aqueous ions **[1]** which conduct electricity. **[1]**

23.7 The covalent radius of sulphur is smaller than that of phosphorus **[1]** hence even 4 chlorines may experience steric crowding. **[1]** It may also be that the S—Cl bond is weaker than the P—Cl bond **[1]** because the steric crowding forces the chlorines away from the sulphur. **[1]**

23.8 (a) (i) SO_2 and SO_3 **[2]**

(ii) SO_2 by burning sulphur in oxygen **[1]** SO_3 by the catalytic oxidation of SO_2 **[1]** with oxygen **[1]** over a vanadium(V) oxide catalyst. **[1]**

(b) $SO_2(g) + H_2O(l) \rightarrow H_2SO_3(aq)$ **[1]** sulphurous acid **[1]**
$SO_3(s) + H_2O(l) \rightarrow H_2SO_4(aq)$ **[1]** sulphuric acid **[1]**

23.9 (a) (i) $Cl_2(g) + H_2O(l) \rightarrow Cl^-(aq) + H^+(aq) + HOCl(aq)$ **[2]**

(ii) $3ClO^-(aq) \rightarrow 2Cl^-(aq) + ClO_3^-(aq)$ **[2]**

(b) ClO^- chlorate(I), linear **[2]**
ClO_3^- chlorate(V), pyramidal **[2]**

23.10 (a) $2F_2(g) + 2H_2O(l) \rightarrow O_2(g) + 4HF(aq)$ **[2]**

(b) The OS of oxygen has increased from -2 to 0 **[1]** representing oxidation **[1]** while that of fluorine has decreased from 0 to -1 **[1]** representing reduction. **[1]**

23.11 Na_2O ionic, MgO ionic, Al_2O_3 ionic, P_4O_{10} covalent, $SO_3(SO_2)$ covalent. **[5]**

23.12 (a) amphoteric **[1]**

(b) K_2O alkaline, BaO basic, Al_2O_3 amphoteric, NO_2 acidic, Cl_2O_7 acidic. **[5]**

23.13 $MgO(s) + 2HCl(aq) \rightarrow MgCl_2(aq) + H_2O(l)$, **[2]** no reaction with alkali. **[1]**
$Al_2O_3(s) + 6HCl(aq) \rightarrow 2Al^{3+}(aq) + 6Cl^-(aq) + 3H_2O(l)$, **[2]**
$Al_2O_3(s) + 2NaOH(aq) \rightarrow 2NaAlO_2(aq) + H_2O(l)$. **[2]**
$SiO_2(s) + 2NaOH(aq) \rightarrow Na_2SiO_3(aq) + H_2O(l)$, **[2]** no reaction with acid. **[1]**

23.14 $NaCl$ no, $MgCl_2$ no, Al_2Cl_6 yes, PCl_3 and PCl_5 yes, SCl_2 and S_2Cl_2 yes. **[8]**

23.15 From ionic to covalent. **[1]** On the LHS of the period are elements with one or two more electrons than an inert gas and they tend to form cations readily and hence ionic oxides and chlorides. **[1]** Near the middle of the period the energetics of the formation of ionic oxides and chlorides becomes unfavourable, as the high ionization energies cannot be offset by lattice enthalpies, **[1]** and the bonding becomes covalent. **[1]**

23.16 Carbon has a strong preference for multiple bond formation **[1]** and hence carbon dioxide is a simple molecular species **[1]** having weak intermolecular forces and a low melting point. **[1]** For silicon, the formation of 4 single Si—O bonds is much more energetically favourable **[1]** than two Si=O bonds (Si rarely forms multiple, i.e. pi-bonds) **[1]** and a giant molecular structure with a high melting point results. **[1]**

23.17 CO_2, $AlCl_3$, SiO_2, HCl. **[4]**

23.18 It suggests that the bonding is covalent (i.e. that the intermolecular forces are weak). **[1]** The Al^{3+} ion has a high charge to size ratio and it polarizes the chloride ion, **[1]** giving its chloride a high degree of covalent character. **[1]**

23.19 $NaCl$ ionic, CCl_4 covalent, PCl_5 covalent, $AlCl_3$ covalent, HCl covalent. **[5]**
$PCl_5(s) + 4H_2O(l) \rightarrow H_3PO_4(aq) + 5HCl(g)(aq)$ **[2]**
$AlCl_3(s) + 3H_2O(l) \rightarrow Al(OH)_3(s) + 3HCl(aq)$ **[2]**
$HCl(g) + H_2O(l) \rightarrow H_3O^+(aq) + Cl^-(aq)$ **[2]**

23.20 The solid is Al_2Cl_6 **[1]** with chlorine bridges which form dative covalent bonds with aluminium **[1]** thus completing its octet of electrons. **[1]** On heating, the chlorine bridging bonds are broken to give monomeric $AlCl_3$. **[1]**

24 The Transition Elements

24.1 (a) Ti [Ar] $3d^2\ 4s^2$, V [Ar] $3d^3\ 4s^2$, Cr [Ar] $3d^5\ 4s^1$, Mn [Ar] $3d^5\ 4s^2$, Cu [Ar] $3d^{10}\ 4s^1$. **[5]**

(b) The full and half full d-orbitals (d^{10} and d^5) are particularly stable, so an electron is transferred from the s- to the d-orbital to achieve these configurations. **[2]**

24.2 $Fe^{2+} = $ [Ar] $3d^6$, $Fe^{3+} = $ [Ar] $3d^5$. **[2]**

24.3 21 (Sc) $= $ [Ar] $3d^1\ 4s^2$, 23 (V) $= $ [Ar] $3d^3\ 4s^2$, 27 (Co) $= $ [Ar] $3d^7\ 4s^2$, 28 (Ni) $= $ [Ar] $3d^8\ 4s^2$, 29 (Cu) $= $ [Ar] $3d^{10}\ 4s^1$, 30 (Zn) $= $ [Ar] $3d^{10}\ 4s^2$. **[6]**

24.4 $Mn^2 = [Ar]\ 3d^5$, $Sc^{3+} = [Ar]$, $Cu^+ = [Ar]\ 3d^{10}$, $Cr^{2+} = [Ar]\ 3d^4$, $Ti^{3+} = [Ar]\ 3d^1$, $Zn^{2+} = [Ar]\ 3d^{10}$. **[6]**

24.5 Mn^{2+} and Fe^{3+}, Cu^+ and Zn^{2+}, Ti^+ and V^{2+}. **[3]**

24.6 **(a)** $+4$, **(b)** $+5$, **(c)** $+7$, **(d)** $+6$, **(e)** $+5$, **(f)** $+4$, **(g)** $+2$, **(h)** $+2$, **(i)** $+3$, **(j)** $+3$, **(k)** $+2$. **[11]**

24.7 **(a)** **(i)** orange/red-brown **(ii)** (pale) green **(iii)** blue **(iv)** colourless **(v)** purple **(vi)** orange. **[6]**

(b) Ligand/water causes splitting of/difference in energy between the d-orbitals **[1]** this energy difference corresponds to a frequency in visible light **[1]** which is absorbed (hence the colour). **[1]**

24.8 complex ion – a transition metal ion (dative covalently) bonded to a number of ligands, ligand – a species capable of donating an electron pair. **[4]**

24.9 **(a)** any three from H_2O, NH_3, $H_2N-CH_2-CH_2-NH_2$, CO etc. **[3]**

(b) any three from F^-. Cl^-, Br^-, I^-, CN^-, SCN^- etc. **[3]**

24.10 **(a)**

Formula
$CuCl_4^{3-}$
$Ag(NH_3)_2^+$
$TiCl_6^{2-}$
$[Co(NH_3)_4(H_2O)_2]^{3+}$
$[Cr(NH_3)_4Cl_2]^+$ **[5]**

(b)

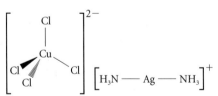

tetrahedral   linear
tetrachlorocuprate(I)   diamminosilver(I)

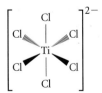

octahedral
(also applies to the Co and Cr complexes)

hexachlorotitanate(IV)
diaquotetramminecobalt(III)
dichlorotetramminechromium(III)

sketches **3 × [1]**
name of shape **3 × [1]**
name of complexes **5 × [1]**

24.11 **(a)** green, (reddish) brown, white, grey/green, green, blue. **[6]**

(b) $Fe(H_2O)_6^{2+}(aq) + 2OH^-(aq) \rightarrow Fe(OH)_2(H_2O)_4(s) + 2H_2O(l)$ **[2]**

(c) $Cr(OH)_3 + 3OH^-(aq) \rightarrow Cr(OH)_6^{3-}(aq)/CrO_3^{3-}(aq) + 3H_2O(l)$ **[3]**

(d) $Ni(OH)_2$, $Co(OH)_2$ and (slightly) $Cr(OH)_3$ **[3]**

24.12 **(a)** **(i)** $CuCl_4^{2-}/CuCl_6^{2-}$ **[1]**
(ii) $Cu(H_2O)_6^{2-}(aq) + 4Cl^-(aq) \rightarrow CuCl_4^{2-}(aq) + 6H_2O(l)$ **[2]**
(iii) ligand displacement **[1]**

(b) **(i)** $Cu(OH)_2$ **[1]**
(ii) $Cu(OH)_2 + 2NH_4^+(aq) \rightarrow Cu(NH_3)_4^{2+}(aq) + 2H_2O(l)$ **[3]**

24.13 **(a)** sulphuric acid manufacture – the Contact Process, ammonia manufacture – the Haber Process, margarine manufacture – hydrogenation. **[3]**

(b) **(i)** $2SO_2 + O_2 \rightleftharpoons 2SO_3$ **[2]**
(ii) $N_2 + 3H_2 \rightleftharpoons 2NH_3$ **[2]**

(c) variable oxidation state **[1]**

25 Shapes of organic molecules

25.1 **(a)** **(i)** 109°, **(ii)** 109°, **(iii)** 120°. **[4]**
(b) **(i)** 109°, **(ii)** 120°, **(iii)** 180°, **(iv)** 120°, **(v)** 109°. **[5]**

25.2 **(a)** **(i)** 109°, **(ii)** 107°.
(b) **(i)** 120°, **(ii)** 120°.
(c) **(i)** 120°, **(ii)** 120°, **(iii)** 107°.
(d) **(i)** 120°, **(ii)** 120°. **[9]**

25.3 **(a)** P, Q, S, R. **[1]**
(b) propanal, propanone, chloropropanoic acid, ethyl ethanoate **[4 + 1 mark for the correct order.]**

25.4 **(a)** **(i)** CH_4 methane

(ii) $CH_3-\overset{\displaystyle CH_3}{\underset{\displaystyle CH_3}{C}}-CH_3$ 2,2-dimethylpropane **[2]**

(b) $\underset{\displaystyle CH_3}{CH_3}C=C\underset{\displaystyle CH_3}{CH_3}$ 2,3-dimethylbut-2-ene **[1]**

(c) $CH_2=CH_2$ ethene **[1]**
(d) $CH\equiv CH$ ethyne **[1]**
(e) **(i)**

and   **[2]**

(ii) Chair because less closeness of hydrogens. **[1]**

(f) (i)

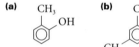

[2]

(ii) There is less steric strain or chlorines are not so close to each other. **[1]**

25.5 (a) A **(b)** A **(c)** A **(d)** E **(e)** A **(f)** D **(g)** E **[7]**

26 Arenes

26.1 (a) (ii) methylbenzene
(ii) ethylbenzene or phenylethane
(iii) 2-methyl-2-phenylbutane
(iv) phenylamine or aminobenzene
(v) benzoic acid
(vi) benzaldehyde
(vii) phenylethanone or acetophenone
(viii) phenol **[8]**

(b) (i), (ii), (iii), (vi), (vii). [5]

26.2

(a)
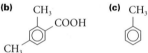

(b)

(c)

(d)

(e)

(f)

[6]

26.3 (a) $\bigcirc$ NO$_2$ nitrobenzene **[2]**
(b) $\bigcirc$ NH$_2$ aminobenzene **[2]**
(c) $\bigcirc$ CH$_3$ methylbenzene **[2]**
(d) $\bigcirc$ COO$^-$ benzoate ion, accept benzoic acid **[2]**
(e) $\bigcirc$ N$_2{}^+$Cl$^-$ benzenediazonium chloride **[2]**

26.4 (a) (i) bromobenzene $\bigcirc$ Br
(ii) chlorobenzene $\bigcirc$ Cl
(iii) phenylethane $\bigcirc$ CH$_2$CH$_3$
(iv) phenylethanone $\bigcirc$ CCH$_3$ ‖ O
(v) 1-phenylpropan-1-one $\bigcirc$ CCH$_2$CH$_3$ ‖ O **[10]**

(b) (i) The ethyl group gives more electron density to the ring than a hydrogen. **[1]** This enhances the 2, 4 and 6 positions. **[1]** The 4 position is less sterically crowded than either the 2 or the 6 positions adjacent to the ethyl group. **[1]** A higher yield of the substituent on the 4 position may be expected than on the adjacent carbons to the ethyl group, but there are two places for substitution so this doubles the opportunity for substitution here. **[1]**

(ii) The nitro group is electron withdrawing compared to hydrogen. **[1]** This makes the benzene ring less susceptible to electrophilic substitution. **[1]** The two positions least effected are the 3 and 5 positions which are the same position with respect to NO$_2$. **[1]** There are two places so two chances of substitution. **[1]**

26.5 (a) Formation of NO$_2{}^+$ from HNO$_3$ and H$_2$SO$_4$. **[1]** Electrons of a double bond from ring attacks NO$_2{}^+$. **[1]** + left on adjacent carbon of ring. **[1]** Electrons in the carbon hydrogen bond on this carbon move to form a new double bond in ring. **[1]** Delocalized ring system reformed. **[1]**

(b) Sn/HCl **[1]**

(c) Nitrous acid or NaNO$_2$/HCl below 4°C **[1]**

(d) (i) $\bigcirc$—N=N—$\bigcirc$—OH

(ii) $\bigcirc$—N=N—$\bigcirc\bigcirc$ HO **[2]**

26.6 (a) Initial attack is by + ion. **[1]** The electron density of the ring is enhanced by electron donating groups. **[1]** Electrons in a double bond from the ring bond with the electrophile. **[1]** The more the electrons are fed into the ring the easier the attack on + ion. **[1]** Also the more electron donation to the ring the more the intermediate + on the ring is spread or stabilized. **[1]**

(b) CH$_3$, OH, NH$_2$. **[1]**

27 Reactions of aldehydes and ketones

27.1 (a) (i)

$$CH_3CH_2\underset{\underset{OH}{|}}{\overset{\overset{CH_3}{|}}{C}}CN$$

(ii)

—NH—N=CCH$_2$CH$_3$ with CH$_3$ substituent and NO$_2$, NO$_2$ on ring

(iii) no reaction

(iv) $CH_3CH_2\underset{\underset{OH}{|}}{CH}{-}CH_3$

(v) $CH_3CH_2\underset{\underset{OH}{|}}{CH}{-}CH_3$ **[6]**

(b) (i) $CH_3CH_2CH_2\underset{\underset{OH}{|}}{CH}{-}CN$

(ii) ring—NH—N=CHCH$_2$CH$_2$CH$_3$ with NO$_2$, NO$_2$

(iii) $CH_3CH_2CH_2COOH$

(iv) $CH_3CH_2CH_2CH_2OH$

(v) $CH_3CH_2CH_2CH_2OH$ **[5]**

27.2 (a) aldehyde, accept CHO **[1]**

(b) Fehling's or Benedicts' solutions **[1]**

(c) ir **[1]**

27.3 (a) reddish precipitate or silver mirror, propanoic acid, CH_3CH_2COOH **[3]**

(e) reddish precipitate or silver mirror, benzoic acid, ⬡COOH **[3]**

(f) reddish precipitate or silver mirror, ethanoic acid, CH_3COOH **[3]**

(b), (c), (d) no changes

27.4 (a) B, A, C **[1]**

(b) R CHO + [H] → R CH$_2$OH **[1]**

(c) Sodium borohydride, NaBH$_4$ **[2]**

(d) LiAlH$_4$ **[1]**

27.5 (a) methanol **[1]**

(b) methanoic acid **[1]**

(c) propan-2-ol **[1]**

(d) 1-phenylethanol **[1]**

(e) phenylmethanol **[1]**

(f) benzoic acid **[1]**

28 Carboxylic acids

28.1 (a) (i) propanoic acid

(ii) methanoic acid

(iii) ethanedioic acid

(iv) benzoic acid

(v) phenylethanoic acid

(vi) 3-hydroxybenzoic acid **[6]**

(b) (i) $CH_3CH_2COOCH_3$

(ii) $HCOOCH_3$

(iii) $\underset{\underset{COOH}{|}}{COOCH_3}$ and $\underset{\underset{COOCH_3}{|}}{COOCH_3}$

(iv) ⬡COOCH$_3$

(v) ⬡CH$_2$COOCH$_3$

(vi) HO⬡COOCH$_3$ **[7]**

28.2 Showing two formulae in equilibrium.

$$-C\overset{\overset{O}{\|}}{\diagdown}_{OH} \rightleftharpoons -C\overset{\overset{OH}{|}}{\diagdown}_{O}$$ **[1]**

Showing curly arrows on one of the formulae. **[1]**

Comment on the tautomers or resonance structures. **[1]**

Hydrogen flips or shared between two.

$$^-C\overset{\overset{O}{/}}{\underset{\underset{O}{\diagdown}}{}}H \quad \text{or} \quad C\overset{\overset{O}{/}}{\underset{\underset{O}{\diagdown}}{}}{-}$$ **[1]**

28.3 (a) $CH_3CH_2COOH + NH_3(aq) \rightleftharpoons$
$CH_3CH_2COO^-NH_4^+ \rightarrow CH_3CH_2CONH_2$ **[3]**

(b) $CH_3CH_2COOH + PCl_3 \rightarrow CH_3CH_2COCl +$
$NH_3(aq) \rightarrow CH_3CH_2CONH_2$ **[3]**

(c) Higher yield because no equilibrium or reversible steps.

28.4 (a) (i) $CH_3COOCH_2CH_3$, ethanoic acid and ethanol **[3]**

(ii) CH_3COOCH_3, ethanoic acid and methanol **[3]**

(iii) $HCOOCH_2CH_3$, methanoic acid and ethanol **[3]**

(b) ⬡COOCH$_2$CH$_3$, benzoic acid and ethanol **[3]**

(c) CH_3CH_2COCl, propanoic acid and hydrochloric acid **[3]**

28.5 Protonation of carboxylic by H$^+$ from concentrated acid. **[1]**

Formation of $CH_3C^+{=}O$ by loss of H$_2$O. **[1]**

Nucleophilic attack by ethanol (lone pair). **[1]**

Release of H^+ from intermediate. **[1]**

Concentrated acid removes H_2O. **[1]**

28.6 $CH_3COOH + CH_3OH \rightleftharpoons CH_3COOCH_3 + H_2O$

[1]

$CH_3COCl + CH_3OH \rightarrow CH_3COOCH_3 + HCl$

[1]

The water or acid is always able to hydrolyse the ester formed even though concentrated acid removes the H_2O and pulls the equilibrium to the right. **[1]** Base neutralizes the HCl forming a salt, or HCl is a better leaving group than water. **[1]**

The neutralization reaction cannot easily be reversed as the chloride is now ionic and is a poor nucleophile to attack the ester. **[1]**

29 Reactions of amines and amides

29.1 (a) (i) $CH_3CH_2CH_2NH_2$

(ii) CH_3CHCH_3
         |
         NH_2

(iii) $CH_3CHCOOH$
          |
          NH_2

(iv) **[4]**

(b)

[2]

(c) (i) $CH_3CHCONH_2$
           |
           NH_2

(ii) $NH_2CH_2CH_2CH_2CH_2CH_2CH_2NH_2$

(iii) $H_2NOCCH_2CH_2CH_2CH_2CONH_2$

(iv) $H_2NOC$$CONH_2$ **[4]**

29.2 (a) $CH_3C{\equiv}N$, ethanenitrile or methyl cyanide **[2]**

(b) $C{\equiv}N$, benzonitrile or phenyl cyanide **[2]**

(c) $N{\equiv}CCH_2CH_2CH_2CH_2C{\equiv}N$,
1,4-dicyanobutane **[2]**

(d) $CH_2CH_2NH_2$, aminoethane **[2]**

(e) $CH_2CH_2NH_2$, 1-amino-2-phenylethane **[2]**

29.3 (a) (i) $CH_3C{\equiv}N \rightarrow CH_3CONH_2 \rightarrow CH_3COOH$ **[2]**

(ii) $C{\equiv}N \rightarrow$ $CONH_2 \rightarrow$ $COOH$ **[2]**

(b) the salt or ion is formed **[1]**

29.4 (a) $NH_2 + NaNO_2/HCl \xrightarrow{<4\,°C}$ $N_2{}^+Cl^-$

$N_2{}^+ + Cl^- + H_2O \xrightarrow{>0\,°C}$ $OH + N_2 + HCl$ **[2]**

(b) $COOH$ **[1]**

(c) (i) $-N{=}N-$$OH$ **[1]**

(ii) 7 **[1]**

29.5 (a)

$CH_3CH_2COOH \xrightarrow{NH_3} CH_3CH_2CO_2{}^- NH_4{}^+ \xrightarrow{heat} CH_3CH_2CONH_2$

$\downarrow PCl_5$ $\qquad\qquad\qquad\qquad\qquad\qquad\qquad\downarrow P_2O_5$

$CH_3CH_2COCl \xrightarrow{NH_3} CH_3CH_2CN$

$\qquad\qquad\qquad\qquad\qquad\qquad\downarrow LiAlH_4$

$CH_3CH_2CH_2NH_2$

alternatively

$CH_3CH_2COOH \xrightarrow{LiAlH_4} CH_3CH_2CH_2OH \xrightarrow{PCl_5} CH_3CH_2CH_2Cl$

$\downarrow NH_3$

$CH_3CH_2CH_2NH_2$

(b) $CH_3CH_2CH_2OH \xrightarrow{PCl_5} CH_3CH_2CH_2Cl \xrightarrow{NH_3} CH_3CH_2CH_2NH_2$ **[2]**

(c) as ions, $CH_3CH_2CH_2NH_3$ and Cl^- **[1]**

(d) (i) $CH_3CHCOOH$
           |
           NH_2 **[1]**

(ii) CH_3CHCOO^-
         |
         $NH_3{}^+$ **[1]**

30 Synthetic pathways

In this chapter, valid alternatives to substances are acceptable.

30.1 (a) A $LiAlH_4$, B PI_3 **[2]**

(b) C $NaBH_4$, D HBr **[2]**

(c) E KOH/C_2H_5OH, F $CH{=}CH_2$, G $HCl(g)$ **[3]**

(d) H KOH/C_2H_5OH, I P_2O_5, J CH_3CHCH_2Br
                                          |
                                          Br **[3]**

30.2 (a) A NaOH, B H_3O^+ **[2]**

(b) C $CH_2CO_2{}^- NH_4{}^+$, D CH_2CO_2H **[2]**

(c) E $Na_2Cr_2O_7/H_3O^+$, F $I_2/NaOH$ **[2]**

(d) G H_3O^+, 320 K, H H_3O^+, 370 K **[2]**

(e) I $KMnO_4/H_3O^+$ and distil out product, J $C_2H_5CO_2H$ **[2]**

30.3 (a) A $KMnO_4/H_3O^+$, B $LiAlH_4$ **[2]**

(b) C HBr, D $OH^-(aq)$, the best reactant is $CaCO_3$ (aq suspension) **[2]**

(c) E $NaNO_2/HCl$, 280 K,   F H_2O, 320 K **[2]**

(d) G CH_3OH/H_2SO_4,

H ⬡CH_2OH,   I CH_3OH **[3]**

30.4 **(a)** A H_3O^+, 370 K,   B CH_3CH_2OH/H_2SO_4 **[2]**

(b) C $Na_2Cr_2O_7/H_3O^+$,   D $CH_3CH_2CO_2H$,

E H_2SO_4,   F $CH_3\ CCH_3$

$\qquad\qquad\qquad\qquad\quad \overset{\|}{O}$ **[4]**

(c) G ⬡$COCl$, H $C_2H_5OH/base$ **[2]**

30.5 **(a)** A H_2SO_4, then H_2O,   B $Na_2Cr_2O_7/H_3O^+$ **[2]**

(b) C OH^-(aq) or $CaCO_3$ (aq suspension),

D $CH_3CH_2CHCH_3$ E $CH_3CH{=}CHCH_3$

$\qquad\qquad\quad \overset{|}{OH}$ **[3]**

(c) F $KMnO_4$, (neutral),   G $Na_2Cr_2O_7/H_3O^+$ **[2]**

30.6 **(a)** A HNO_3/H_2SO_4,   B Fe/H_3O^+ **[2]**

(b) C KCN,   D $LiAlH_4$ **[2]**

(c) E $CH_3CH_2CONHCH_3$,   F $LiAlH_4$ **[2]**

(d) G NH_3(aq),   H $LiAlH_4$ **[2]**

30.7 A CH_3CHCH_3 B CH_3CHCH_3

$\qquad\quad \overset{|}{Br}$ $\qquad\quad \overset{|}{CN}$

C CH_3

$\quad\searrow$

$\qquad CHCO_2CH_2CH_2CH_2CH_3$

$\quad\nearrow$

$\ \ CH_3$

D CH_3 E $CH_3CH_2CH_2CH_2OH$

$\quad\searrow$

$\qquad CHCO_2{}^-Na^+$

$\quad\nearrow$

$\ \ CH_3$

F $CH_3CH_2CH_2CO_2H$ G CH_3

$\qquad\qquad\qquad\qquad\quad\searrow$

X PCl_5 **[8]** $\qquad CHCO_2H$

$\qquad\qquad\qquad\qquad\quad\nearrow$

$\qquad\qquad\qquad\quad\ \ CH_3$

30.8 A CH_3CHCH_3,   B CH_3CCH_3

$\qquad\quad \overset{|}{OH}$ $\qquad\quad \overset{\|}{O}$

C CH_3CHCH_3

$\qquad\quad \overset{|}{Br}$ **[3]**

30.9 A $CH_3CH{=}CH_2$ B CH_3CHCH_3

C CH_3CCH_3 $\qquad\qquad \overset{|}{OH}$

$\qquad \overset{\|}{O}$

D CH_3 E $CH_3CO_2{}^-Na^+$

$\quad\searrow\qquad\qquad NO_2$

$\qquad C{=}NNH$⬡NO_2

$\quad\nearrow$

$\ \ CH_3$

F CHI_3 G CH_3CO_2H **[7]**

30.10 A ⬡NO_2 B ⬡NH_2

C ⬡$N_2{}^+$ D ⬡OH **[4]**

30.11 A ⬡NO_2 B ⬡NH_2

C ⬡$N_2{}^+$ D ⬡CO_2H **[4]**

30.12 A ⬡$CH{=}CH_2$ B ⬡$CHCH_2Br$

C ⬡$CHCH_3$ $\qquad\qquad\quad \overset{|}{Br}$

$\quad \overset{|}{Br}$ **[3]**

30.13 A $\overset{\displaystyle\ }{+}CH{-}CH_2\overset{\displaystyle\ }{+}$ B O_2N⬡CH_2CH_2Br

$\qquad \overset{|}{\underset{\displaystyle ⬡}{}}$

$\qquad\ \ NO_2\ \big]_n$

C $^+NH_3$⬡CH_2CH_2Br D $^+N_2$⬡CH_2CH_2Br

E $BrCH_2CH_2$⬡$N{=}N$

$\qquad\qquad\qquad\qquad \overset{|}{⬡}$

$\qquad\qquad\qquad\qquad OH$

F HO⬡CH_2CH_2Br X P_2O_5 **[7]**

30.14 A $HO_2CCH_2CH_2Br$, B $CH_3CH_2O_2CCH_2CH_2Br$,

C $HOCH_2CH_2CH_2CN$, D $HOCH_2CH_2CH_2CO_2H$,

E $CH_3CO_2CH_2CH_2CH_2Br$ **[5]**

31 Polymerization

31.1 **(a)**

polypropene	addition
terylene	condensation polyester
nylon	condensation polyamide
acrylic	addition
perspex	addition
protein	condensation polyamide
polythene	addition
rubber	addition
PVA	addition
PTFE	addition
PVC	addition
polystyrene	addition **[12]**

(b) (i) $\quad CH_3$ **(ii)** ⬡

$\qquad\ \ +CH{-}CH_2+$ $\qquad +CH{-}CH_2+$

(iii) $+CH_2{-}CH_2+$

(iv) $\qquad\qquad\ \ H\qquad\quad\ H$

$\qquad\qquad\qquad\ \overset{|}{}\qquad\qquad \overset{|}{}$

$+C(CH_2)_4C{-}N{-}(CH_2)_6{-}N+$

$\ \ \overset{\|}{O}\qquad\ \overset{\|}{O}$

(v) $\qquad\qquad\quad H$

$\qquad\qquad\qquad \overset{|}{}$

$+C(CH_2)_4C{-}N+$

$\ \overset{\|}{O}$

(vi) $+C$⬡$C{-}O{-}CH_2CH_2O+$

$\quad\ \overset{\|}{O}\qquad\quad \overset{\|}{O}$ **[6]**

31.2 **(a)** **(i)** They are both polyamides. **[1]** Amides are hydrolysed by acids to carboxylic acids and the salt of the amine or ammonia. **[1]**

(ii) It is a polyester. **[1]** Polyesters are hydrolysed by alkali to form salts of carboxylic acids and alcohols. **[1]**

(iii) Polythene is an addition polymer – it is a large hydrocarbon. Hydrocarbons are unreactive toward acid or base. **[1]**

(b) 6-aminohexanoic acid, $HOOC(CH_2)_5NH_2$ **[2]**

31.3 **(a)** Atactic, as it does not require special catalysts or conditions. **[1]**

(b) atactic – soft or pliable items **[1]** such as squeezy toys. **[1]**
isotactic – firm objects which are not rigid **[1]** such as buckets or rope. **[1]**
syndiotactic – stiff objects **[1]** such as bowls in sinks or casings. **[1]**

(c)

$$+C-C-C-C-+$$

with CH_3 groups

The alternate side chains **[1]** on every other carbon along the chain. **[1]**

(d) Homolytic cleavage of one of the bonds in the double bond and attack by the free radical on another double bond. **[1]**
Cleavage of this double bond to form a further free radical to continue the process. **[1]**

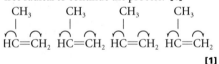

[1]

32 Testing for functional groups

32.1 **(a)** **(i)** primary alcohol **[1]**

(ii) aldehyde **[1]**

(b) gives off hydrogen
forms a salt, a sodium alcoxide **[2]**

(c) carboxylic acid **[1]**

32.2 **(a)** $X = CH_3CH_2OH$
$Y = CH_3COOH$ **[2]**

(b) $CH_3COOH + CH_3CH_2OH \rightarrow$
$CH_3COOCH_2CH_3 + H_2O$
[1 mark for each product.]

(c) Forming an ester with an acid and an alcohol is an acid catalysed equilibrium. **[1]** Using the acid chloride the reaction goes to completion. **[1]**

32.3 **(a)** C_4H_7Br **[1]**

(b) CH_2＝$CHCH_2CH_2Br$ **[1]**
CH_3CH＝$CHCH_2Br$ *cis/trans* **[2]**
CH_3CH_2CH＝$CHBr$ *cis/trans* **[2]**
$CH_3CHBrCH$＝CH_2 optical **[2]**
CH_3CBr＝$CHCH_3$ *cis/trans* **[2]**
CH_2＝$CBrCH_2CH_3$ **[1]**

$$CH_2=CCH_2Br \text{ with } CH_3 \text{ [1]}$$

$$BrCH=CCH_3 \text{ with } CH_3 \text{ [1]}$$

32.4 **(a)** benzene-$COOH$ **[1]**

(b) **(i)** benzene-$COCl$ or benzoyl chloride **[1]**

(ii) benzene-COO^- or benzoate ion **[1]**

(c) It is an aromatic compound or has a benzene ring. **[1]**

32.5 **(a)** CH_2OH
CH_2OH **[1]**

(b) 62 **[1]**

(c) **(i)** no reaction **[1]**   **(ii)** no reaction **[1]**

(iii) forms CH_2Cl
CH_2Cl **[1]**

(iv) forms $COOH$
$COOH$ **[1]**

32.6 **(i)** C＝O or a carbonyl

(ii) CHO or an aldehyde

(iii) C＝O or a ketone

(iv) acetaldehyde only

(v) methyl ketone

(vi) ketone that has no methyl group adjacent to the carbonyl function

(vii) $COOH$ or a carboxylic acid

(viii) C＝C or an alkene

(ix) chloride or chloroalkane

(x) bromide or bromoalkane

(xi) iodide or iodoalkane **[11]**

33 Modern analytical techniques

33.1 **(a)** **(i)** Chemical shifts cannot be quoted absolutely because they depend on the field strength and RF frequency of the spectrometer **[1]** so an internal standard of agreed chemical shift is needed against which the chemical shift can be measured. **[1]**

(ii) tetramethylsilane (TMS) $(CH_3)_4Si$ **[1]**

(iii) It has a relatively large number of protons **[1]** so not much need be added to be detected. **[1]** All the protons in it are chemically identical **[1]** so it gives a single peak in the NMR. **[1]** It is readily soluble in most organic liquids. **[1]** It is relatively unreactive. **[1]**

(iv) zero ppm on the δ scale **[1]**

(b) (i) It contains no hydrogen, resonances from which may obscure peaks in the sample. **[1]** It is a good solvent for many organic substances. **[1]**

(ii) deuterated water, D_2O **[1]**

33.2 (a)

substance	NMR evidence for structure	IR evidence for structure
$CH_3—O—CH_3$	single line in spectrum **[1]**	no O—H stretch **[1]**
C_2H_5OH	characteristic splitting pattern of the ethyl group: a triplet around 1δ (CH_3) and a quartet around 3.5δ ($—CH_2—$)/ppm **[2]**	strong (broad) absorption due to O—H at about $3300\ cm^{-1}$ **[1]**

(b) Only ethanol can give rise to a peak of m/e 29 in the mass spectrum. **[1]**

33.3 Reduction converts C=O to $—CH_2—OH$ **[1]** so *either* the reduction in the intensity of the C=O stretch at $1700–1720\ cm^{-1}$ *or* the appearance of a strong absorption at about $3300\ cm^{-1}$ due to O—H. **[1]**

33.4 (a) (i) NMR spectrum of 1,1,2-trichloroethane:

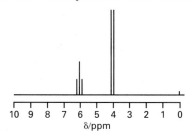

[4]

(ii) Because the coupling constants for the doublet and triplet are identical. **[1]**

(b) (i) Mass spectrum of 1,1,2-trichloroethane:

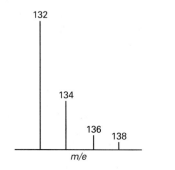

[3]

(ii) The structural isomer $CH_3–CCl_3$ can lose a methyl group **[1]** leading to a M-15 peak in the mass spectrum lacking in the other isomer. **[1]**

33.5 $1750\ cm^{-1}$ absorption corresponds to C=O, $1250\ cm^{-1}$ absorption corresponds to C—O of ester group, **[2]** the parent ion peak gives the RMM as 88, which fits ethyl ethanoate. **[1]**

the peak at 43 (M-29) corresponds to loss of $C_2H_5^+$, **[1]**

the peak at 73 (M-15) corresponds to loss of CH_3^+, **[1]**

the NMR confirms an ethyl group (quartet at 4δ, triplet at 1δ/ppm), **[1]**

the single peak at 2δ/ppm being due to the CH_3 group of the parent acid, **[1]**

the relative intensities are consistent with these assignments, **[1]**

X is ethyl ethanoate, **[1]** an ester. **[1]**

33.6 (a) Infra-red would be least useful in distinguishing the two isomers **[1]** because they contain the same types of bonds **[1]** and hence would give similar IR spectra. **[1]**

(b) Difficult to say whether mass spectra or NMR would be more useful in distinguishing the two isomers (allow answer supporting either) i.e. mass spectra because 1-chloro-4-methylbenzene would have a peak at (M-15) due to loss of CH_3^+ **[1]** which cannot occur with (chloromethyl)benzene **[1]** *or* NMR because the integration would show 4 aromatic protons to 3 aliphatic ones in 1-chloro-4-methylbenzene, **[1]** but a 5:2 ratio for (chloromethyl)benzene. **[1]**

33.7 (a) blue **[2]**

(b) blue **[2]**

(c) red/purple **[2]**

33.8 (a) quartet **[2]**

(b) triplet **[2]**

(c) singlet **[2]**

33.9 (a) C/H analysis gives C_7H_8 **[1]** high C:H ratio suggests an unsaturated molecule **[1]** IR consistent with arene (C—C, C—H stretching and also the strong absorption at around $720\ cm^{-1}$ support the existence of a benzene ring) **[2]**

the parent ion in the MS supports the RMM of 92 **[1]**

the chemical shift in the NMR supports a

phenyl group (5 protons) and a methyl side-chain (3 protons) **[2]**
X is methylbenzene **[1]**

(b) The product would be benzoic acid **[1]** so expect an O—H stretch at 3000–3300 cm^{-1} **[1]** and a C=O stretch at around 1700 cm^{-1}. **[1]**

(c) The lack of m/e 15 suggests that the CH$_3$ might be part of the fragment corresponding to m/e 91 **[1]** m/e 91 corresponds to C$_7$H$_7^+$ so this assumption seems justified **[1]** the m/e 91 fragment is in fact the cyclic C$_7$H$_7^+$ **[1]**

33.10 Elemental analysis gives C$_2$H$_3$ClO **[1]** RMM = 80, based on ^{37}Cl **[1]** MS peak at m/e 78 consistent with presence of a single chlorine atom **[1]**, peaks at m/e 43 (CH$_3$CO$^+$) **[1]** and 15 (CH$_3^+$) **[1]** point to ethanoyl chloride, IR indicates C=O (1800 cm^{-1}) **[1]** and C—Cl (600 cm^{-1}) **[1]** NMR consistent with presence of a methyl group – no splitting of the resonance. **[1]**
The reaction of A with water is consistent with its being ethanoyl chloride **[1]**.

33.11 Elemental analysis B gives C$_4$H$_8$CO **[1]** consistent with m/e 72 as parent ion peak in MS **[1]** the absorption at 1700 in the IR suggests C=O, **[1]** the MS peak at m/e 43 (CH$_3$CO$^+$) is most informative, suggesting a methyl ketone, **[1]** other peaks supporting butanone as the structure **[1]** are m/e 29 (C$_2$H$_5^+$) **[1]** and m/e 57 C$_2$H$_5$CO$^+$. **[1]**
The NMR data are consistent with this formulation: **[3]**

chemical shift (δ) /ppm	multiplicity	relative intensity	assignment
2.5	quartet	2	CH$_3$COCH$_2$CH$_3$
2.1	singlet	3	CH$_3$COCH$_2$CH$_3$
1.0	triplet	3	CH$_3$COCH$_2$CH$_3$

(a) butanone (see above) **[1]**

(b) add iodine in potassium iodide solution, **[1]** aqueous sodium hydroxide, **[1]** warm, **[1]** a yellow ppt. of iodoform (triiodomethane) CHI$_3$ **[1]** confirms a methyl ketone. **[1]**

(c) (i) CH$_3$CH$_2$CH$_2$CHO **[1]** butanal. **[1]**
(ii) add ammoniacal silver nitrate/Tollens' reagent **[1]** a silver mirror forms with butanal only. **[1]**

33.12 IR data suggests an aldehyde: C=O at 1730 cm^{-1} **[1]** C—H stretch in an aldehyde at 2700 and 2800 cm^{-1} **[1]** MS supports propanal, if m/e 58 is the parent ion (likely) **[1]**, m/e 57 consistent with

loss of hydrogen from —CHO **[1]**, m/e 29 accounted for by C$_2$H$_5^+$ **[1]**
The NMR supports the propanal structure:
9.8δ/ppm (singlet, 1 proton) CH$_3$CH$_2$CHO **[1]**;
2.5δ/ppm (quartet, 2 protons) CH$_3$CH$_2$CHO **[1]**
at higher resolution each peak in the quartet appears as a doublet as a result of further coupling with the —CHO proton **[1]** 1.1δ/ppm (triplet, 3 protons) —CH$_3$CH$_2$CHO. **[1]**

33.13 (a) Elemental analysis gives C$_3$H$_7$NO as the EF, **[1]** that this is also the MF is supported by parent ion peak m/e 73. **[1]** IR suggests —NH$_2$ group (3340 and 3200 cm^{-1}) **[1]** and a C=O (1640 cm^{-1}) **[1]** which, together with the liberation of ammonia when A is heated with alkali, confirms an amide group —CONH$_2$. **[1]** The reaction of A with phosphorus(V) oxide to give a compound of molecular formula C$_3$H$_5$N involves dehydration of the amide group to give propanonitrile, C$_2$H$_5$CN. **[1]**
The NMR data confirm A as propanamide, C$_2$H$_5$CONH$_2$. **[1]**

NMR spectrum /ppm	6.2δ (broad line, 2 protons), CH$_3$CH$_2$CONH$_2$ **[1]**
	2.2δ (quartet, 2 protons), CH$_3$CH$_2$CONH$_2$ **[1]**
	1.1δ (triplet, 3 protons), CH$_3$CH$_2$CONH$_2$ **[1]**

(b) heat under reflux **[1]** with bromine **[1]** in aqueous sodium hydroxide solution **[1]**

(c) basicity depends on availability of nitrogen lone pair, **[1]** in ethanamide the (electronegative) oxygen withdraws electron density **[1]** making the lone pair less basic **[1]**

33.14 Elemental analysis gives C$_4$H$_7$O$_2$Br as the EF **[1]** confirmed as the MF by parent ion at m/e 168 and 166 **[1]** and two peaks in this region further confirm the presence of a single Br in the molecule **[1]** the intense peak at 87 corresponds to a loss of bromine **[1]**. IR suggests an ester (1740 and 1200 cm^{-1}) **[1]** and the NMR suggests two pairs of non-equivalent protons **[1]** and is consistent with the formulation of B as methyl-3-bromopropanoate **[1]**:

NMR spectrum /ppm	3.7δ (singlet, 3 protons), CH$_3$—O—CO—CH$_2$CH$_2$—Br **[1]**
	3.5δ (triplet, 2 protons), CH$_3$—O—CO—CH$_2$CH$_2$—Br **[1]**
	2.9δ (triplet, 2 protons), CH$_3$—O—CO—CH$_2$CH$_2$—Br **[1]**

33.15 **(a)** Oxidation of C to benzoic acid confirms the presence of C_6H_5- **[1]** the side chain on the benzene ring must be C_2H_5 ($106 - 77 = 29$). **[1]** (The origin of m/e 91 peak is discussed separately in **(c)**.) The NMR data are consistent with C being ethylbenzene **[1]**:

NMR spectrum /ppm	
7.1δ (singlet, 5 protons), $C_6H_5-CH_2CH_3$ **[1]**	
2.7δ (quartet, 2 protons), $C_6H_5-CH_2CH_3$ **[1]**	
1.2δ (triplet, 3 protons), $C_6H_5-CH_2CH_3$ **[1]**	

(b) aluminium chloride (catalyst), **[1]** chloroethane **[1]** anhydrous conditions **[1]** heat under reflux **[1]**

(c) the very intense peak at m/e 91 in the mass spectrum corresponds to $C_7H_7^+$ **[1]** it might have been formed by loss of CH_3, **[1]** to give $C_6H_5-CH_2^+$, which then undergoes ring opening followed by formation of seven-membered $C_7H_7^+$ ring **[1]**

33.16 In addition to the parent ion peak at m/e 130, the following information is most useful: m/e 115 (M-15) indicating loss of CH_3 **[1]**; m/e 87 (M-43) indicating loss of CH_3CO **[1]**; m/e 85 (M-45) indicating loss of CH_3CH_2O. **[1]** Taken along with the IR evidence this suggests the structure $CH_3COCH_2COOCH_2CH_3$. This structure is consistent with NMR data:

NMR spectrum /ppm	
4.1δ (quartet, 2 protons) $CH_3COCH_2COOCH_2CH_3$ **[1]**	
3.3δ (singlet, 2 protons) $CH_3COCH_2COOCH_2CH_3$ **[1]**	
2.2δ (singlet, 3 protons) $CH_3COCH_2COOCH_2CH_3$ **[1]**	
1.2δ (triplet, 3 protons) $CH_3COCH_2COOCH_2CH_3$ **[1]**	

E is thus ethyl-3-oxobutanoate (ethylacetoacetate). **[1]**

Part 3
Synoptic questions

1 (a) Silicon(IV) oxide is a giant covalent substance in which boiling involves the breaking of strong covalent bonds, requiring much energy, hence the high boiling point of 2230 °C. **[1]** Carbon(IV) oxide (carbon dioxide) is a simple molecular substance with weak van der Waals' intermolecular bonds, hence the much lower boiling point of −78 °C. **[1]** The preference of silicon to form four single bonds (and hence a giant covalent structure), rather than two double bonds **[1]** can be traced to the decreased ease of overlap of its 3p orbitals, compared to the 2p in carbon, to form π-bonds. **[1]**

(b) Copper has more valence electrons than potassium **[1]** and hence is a better conductor because there are more potential charge carriers. **[1]** The lithium atom is much smaller than that of caesium and its valence electrons are closer to the nucleus **[1]** and experience a greater force from it, so the metallic bonds are stronger and so lithium is much harder than caesium. **[1]**

(c) Water is polar **[1]** and so the solvation enthalpies of ions in aqueous solution are highly exothermic and able to offset the energy needed to break down the crystal lattice. **[1]** The solvation enthalpies in the non-polar tetrachloromethane are considerably smaller than the lattice energy, so solution does not occur. **[1]** Iodine dissolves in tetrachloro-methane largely as a result of entropy since the transition from the relative order of the molecular iodine crystal to the random arrangement of the dissolved molecules involves a considerable (favourable) entropy increase. **[1]** In water there exist strong intermolecular hydrogen bonds **[1]** which would have to be disrupted when iodine dissolved in water and replaced by intermolecular forces between the water and iodine of magnitude comparable to the hydrogen bond itself. **[1]** Iodine is non-polar and the intermolecular forces between it and water are too weak to achieve this. **[1]**

(d) Hydrogen bonds exist in ammonia **[1]** which are considerably stronger than the van der Waals' forces which are the dominant intermolecular forces in the other two hydrides, **[1]** hence the much higher boiling point of ammonia. The arsenic atom has more electrons than phosphorus **[1]** and so the intermolecular van der Waals' forces are stronger and the boiling point higher than that of phosphine. **[1]**

(e) Diamond is a giant covalent substance having all its electrons involved in covalent bonding and unable to move to carry an electric current. **[1]** The highly-connected three-dimensional structure is very hard because much force is needed to move one atom with respect to another. **[1]**
In graphite only three of the electrons on carbon are used in bonding to give a layer structure held together by weak van der Waals' forces, allowing one layer to slide easily over another. **[1]** The fourth electron from each carbon atom is accommodated between the layers of atoms as a mobile flux of electrons, which enables a current to flow. **[1]**

2 (a) (i)

halide	product containing a halogen	product containing sulphur
NaBr	bromine + hydrogen bromide **[1]**	sulphur dioxide **[1]**
NaI	iodine + traces of hydrogen iodide **[1]**	mixture of products including hydrogen sulphide, sulphur and sulphur dioxide **[1]**

(ii) Iodide ions are sufficiently powerful reducing agents **[1]** to reduce the sulphate ion to hydrogen sulphide, **[1]**

while the less powerful reducing agents, bromide ions, reduce it only to sulphur dioxide. **[1]** Iodide is a better reducing agent because its outer electron is more easily lost, being further from the nucleus and better shielded from it. **[1]**

(b)

halide ion	observation with aqueous silver(I) ions	observation when aqueous ammonia is added to halide plus silver ions
Cl^-	white precipitate formed **[1]**	precipitate dissolves completely **[1]**
Br^-	pale-yellow precipitate **[1]**	precipitate dissolves partially **[1]**
I^-	yellow precipitate **[1]**	precipitate totally insoluble **[1]**

(c) (i) $3ClO^-(aq) \rightleftharpoons ClO_3^-(aq) + 2Cl^-(aq)$ **[2]**

(ii) Disproportionation is a reaction in which an element in a particular species is simultaneously oxidised and reduced. One of the chlorines in ClO^- has increased in OS from $+1$ to $+5$ and so has been oxidized, **[1]** while two of the chlorines in ClO^- have decreased in OS from $+1$ to -1 and so have been reduced. **[2]**

(d) (i) When halogens behave as oxidizing agents, they are reduced in the process, and this involves formation of the corresponding halide ion. **[1]** Chlorine is a more powerful oxidizing agent than bromine because formation of the chloride ion is more energetically favourable than formation of the bromide ion **[1]** because the chlorine atom is smaller **[1]** and the added electron is attracted more strongly. **[1]** (A factor which should work in favour of bromine as an oxidizing agent, namely the weaker Br—Br bond (193 kJ ml^{-1} compared to 242 kJ mol^{-1} for Cl—Cl) must be unable to offset the greater electron affinity and hydration enthalpy of chlorine.)

(ii) Aqueous solutions of hydrogen fluoride are only weakly acidic because hydrogen

fluoride has strong intermolecular hydrogen bonds, **[1]** which must be broken before ionization of individual molecules can occur. **[1]** The very strong H—F bond **[1]** means that the energetics of ionization, in which the bond breaking is compensated by the hydration enthalpies of the resulting aqueous ions, lies in favour of the undissociated hydrogen fluoride molecules. **[1]**

3 (a) (i) The $E^\ominus$ value of the standard hydrogen electrode $= 0.00$ V **[1]**

(ii) Standard conditions are: temperature 298 K, solution concentrations 1.00 M, gas pressure 1 atmosphere **[3]**

(b) (i) Fe **[1]**

(ii) Cl_2 **[1]**

(iii) Br_2 and Cl_2 **[2]**

(c) (i) positive **[1]**

(ii) The conditions might not be standard (most commonly the temperature is not 298 K and solutions are not 1.00 M) **[1]** the activation energy of the reaction may be large **[1]**

(d) (i) $Fe(s) \mid Fe^{2+}(aq) \parallel Cu^{2+}(aq) \mid Cu(s)$ **[2]**

(ii) $+0.78$ V **[2]**

(e) (i) Reduction occurs at the copper electrode **[1]**

(ii) $Cu^{2+}(aq) + 2e^- \rightarrow Cu(s)$ **[2]**

(iii) $Cu^{2+}(aq) + Fe(s) \rightarrow Cu(s) + Fe^{2+}(aq)$ **[2]**

4 (a) (i) RAM of M $= 39$, M likely to be potassium **[3]**

(ii) $1s^2\ 2s^2\ 2p^6\ 3s^2\ 3p^6\ 4s^1$ **[1]** The ionization energy of the 4s electron is sufficiently small to be offset by the energy released on compound formation or when aquated ions result. **[1]** There is no bar to the loss of further (3p) electrons, though no compounds of potassium in other than the $+1$ oxidation state are known, **[1]** no doubt because the ionization of electrons from the complete 3p orbital involves too much energy. **[1]**

(b) $K(g) - e \rightarrow K^+(g)$ **[2]**

(c) (i) Hydrogen gas evolved **[1]** which ignites with the heat of the reaction and burns with a lilac flame **[1]** (due to the potassium ions present), **[1]** metal floats on the water. **[1]**

$$2K(s) + 2H_2O(l) \rightarrow 2KOH(aq) + H_2(g)$$
[1]

(ii) The element below M has a larger atomic radius **[1]** and more intervening electron shells to shield its single outer electron, **[1]** which is lost in forming the $+1$ ion when it reacts with water. **[1]** The easier loss of the electron is the main factor in contributing to the general increase in reactivity with increase in atomic number within a group. **[1]**

(d) (i) The empirical formula of the oxide Y is MO. **[1]** The group 1 ion must be M^+, **[1]** so a likely formulation for an empirical formula MO is M_2O_2, containing the O_2^{2-} (peroxide) ion. **[1]**

(ii) It is always difficult to argue the reasons for the non-existence of a compound, but a possible reason might be that the small lithium ion **[1]** would result in a rather small lattice enthalpy for Li_2O_2 compared to the simple oxide Li_2O, given the significantly larger ionic radius of O_2^{2-} (0.180 nm) compared to O^{2-} (0.140 nm). **[1]**

5 (a) (i) protons $= 12$, neutrons $= 13$, electrons $= 12$ **[2]**

(ii) 24.3 **[2]**

(iii)

abundance

axes **[1]**
correct m/e **[1]**
correct relative intensities **[1]**

(b) (i) $M(g) - e \rightarrow M^+(g)$ **[2]**

(ii) The atomic radius increases, placing the electron that is lost further from the nucleus, **[1]** and there are more intervening shells to shield the outer

electron from the nuclear charge. **[1]** These effects combine to decrease the ionization energy as atomic number increases within the group.

(iii) The nuclear charge increases by one unit as each successive electron is added **[1]** but the added electron is not shielded fully from the increased attraction of the nucleus, **[1]** hence the ionization energy increases as the atomic number increases on crossing a period.

(c) (i) The electronic configuration of the two atoms are B $1s^2 2s^2 2p^1$ and Be $1s^2 2s^2$. The electron from B comes from the 2p orbital which is further from the nucleus **[1]** (and is more shielded by inner shells), while that lost from Be is in a 2s orbital which is closer to the nucleus and more influenced by nuclear charge. **[1]**

(ii) The electron lost from oxygen is paired with a second electron in one of the 2p orbitals and experiences electrostatic repulsion as a result and is therefore easier to ionize. **[1]** There are no paired electrons in the 2p orbitals in nitrogen; each electron occupies a 2p orbital alone and so there is no repulsion. **[1]**

6 (a) (i) Ba, **(ii)** Ba, **(iii)** Mg, **(iv)** Ba. **[4]**

(b) Increasing cation size leads to an increase in the thermal stability of the carbonates of Group 2. **[1]** One possible explanation stresses the importance of comparable cation and anion radius if a crystal is to have a high lattice enthalpy and since the carbonate ion is significantly larger than the oxide ion, **[1]** the carbonates of larger cations will be most resistant to thermal decomposition. **[1]** Another explanation is that smaller cations will tend to polarize the carbonate ion to a greater extent than larger ones **[1]** which will facilitate the breaking of the C—O bond required in thermal decomposition:

$$MCO_3 \rightarrow MO + CO_2. \text{ [1]}$$

(c) (i) The RAM of M is 88, so M is Sr **[4]**

(ii) 0.208 g **[2]**

7 (a) Na_2O – alkaline, SO_2/SO_3 – acidic, Al_2O_3 – amphoteric. **[3]**

(b) (i) Ionic – sodium/magnesium/aluminium, simple molecular – phosphorus/ sulphur/ chlorine, giant covalent – silicon. **[3]**

(ii) Difference – ionic substances conduct electricity if molten or in solution [1] because they contain ions which, when mobile, can carry a current. [1] No mobile charge carriers (ions or electrons) exist in giant covalent structures. [1]

(iii) Similarity – ionic and giant molecular substances both have high melting/boiling points. [1] The forces of attraction between the ions in an ionic substance and the strengths of the covalent bonds in a giant molecular substance are similar [1] and both are strong. [1]

8 (a) X: $2Cl^- - 2e \rightarrow Cl_2$; Y: $2H^+ + 2e^- \rightarrow H_2$ [2]

(b) (i) cathode

(ii) The solution contains the following ions: Na^+, H^+, Cl^- and OH^-. [1] As the electrolysis proceeds, H^+ ions are removed irreversibly from the cathode compartment as hydrogen gas, [1] driving the ionization of water to the RHS: $H_2O(l) \rightleftharpoons H^+(aq) + OH^-(aq)$ [1] and leading eventually to the formation of aqueous sodium hydroxide solution, contaminated with NaCl. [1]

(c) (i) $OH^-(aq) + Cl_2(aq) \rightleftharpoons Cl^-(aq) + ClO^-(aq) + H^+(aq)$ [2]

(ii) $3ClO^-(aq) \rightleftharpoons ClO^-_3(aq) + 2Cl^-(aq)$ [2] One of the chlorines in ClO^- has increased in OS from +1 to +5 and so has been oxidized, [1] while two of the chlorines in ClO^- has decreased in OS from +1 to −1 and so has been reduced. [1]

(d) (i) The sodium chloride must be molten because in the presence of water the discharge of hydrogen ions ($2H^+ + 2e^- \rightarrow H_2$) occurs in preference to the discharge of sodium ions ($Na^+ + e^- \rightarrow Na$) [1] and no sodium metal can be obtained. [1]

(ii) sodium – coolant in nuclear plant/manufacture of tetraethyllead/titanium manufacture chlorine – manufacture of chlorinated solvents/water purification/plastics manufacture [2]

9 (a) (i) $MnO_4^- + 5Fe^{2+} + 8H^+ \rightarrow Mn^{2+} + 5Fe^{3+} + 4H_2O$ [3]

(ii) 9.80 g [2]

(b) $MnO_4^- + 8H^+ + 5e^- \rightarrow Mn^{2+} + 4H_2O$
$NO_2^- + H_2O - 2e^- \rightarrow NO_3^- + 2H^+$ [4]

(c) (i) By the first pale pink colour of excess manganate(VII) in the solution. [1]

(ii) 0.05 mol dm^{-3} [2]

(d) (i) A single species which oxidizes itself and is simultaneously reduced. [2]

(ii) $2Mn^{3+}(aq) + 2H_2O(l) \rightleftharpoons MnO_2(s) + Mn^{2+}(aq) + 4H^+(aq)$ [3]

(iii) Increasing the pH increases the tendency of Mn^{3+} to disproportionate [1] because the position of the above equilibrium would be displaced to the RHS in an attempt to restore the hydrogen ion concentration. [1] Reducing the pH has the opposite effect.

10 (a) (i) $Al(Al_2O_3)$,

(ii) $Cl(Cl_2O$ and $Cl_2O_7)$,

(iii) $C(CO_2)$ only. [3]

(b) (i) sodium, magnesium or aluminium oxide

(ii) silicon(IV) oxide [dioxide]

(iii) phosphorus(V) oxide, sulphur di- or trioxide, chlorine di- or heptoxide [3]

(c) Similarity: ionic and giant molecular substances have high melting/boiling points. Difference: ionic substances conduct when molten or in solution, giant molecular substances do not. [2]

11 (a) The enthalpy change associated with the removal of one mole of electrons from a mole of gaseous atoms of the element. [2]

(b) (i) $Mg(g) - e \rightarrow Mg^+(g)$ [1]

(ii) $Mg^+(g) - e \rightarrow Mg^{2+}(g)$ [1]

(c) The first electron is removed from an uncharged atom, [1] while the second is removed from a singly charged cation. [1] The attractive forces between an electron and a cation are stronger than those between an electron and a neutral atom, hence more energy is required to remove the electron. [1]

(d) (i) Across the period from Li to Ne there is a gradual increase in nuclear charge, [1] from which the electron added at the same time is not well shielded. [1] So there is a general increase in first ionization energy from Li to Ne as the attraction from the nucleus increases. [1]

(ii) The electronic configuration of the oxygen atom is $1s^2 2s^2 2p^4$ so one of the electrons in the 2p orbital must be paired **[1]** and so experiences electrostatic repulsion from the other electron in the same orbital. **[1]** No such repulsion exists in nitrogen ($1s^2 2s^2 2p^3$) because the three p electrons occupy the 2p orbitals singly without pairing, **[1]** so the first ionization energy of oxygen is lower than that of nitrogen. **[1]**

(e) 63.6 **[2]**

(f) **(i)** m/e 93 and 95 **[2]**

(ii) $CuCH_3^+$ $m/e = 78$ (or 80), CH_3^+ $m/e = 15$. **[2]**

12 (a) Ethene is a simple molecular substance **[1]** having weak van der Waals' intermolecular forces **[1]** while polyethene is macromolecular **[1]** so the van der Waals' forces between the polymer chains are very much greater than those in molecular ethene. **[1]** Polyethene therefore melts at a higher temperature.

(b) In diamond all of the electrons are involved in covalent bonding **[1]** to create a giant molecular structure based on four bonds to each carbon atom. **[1]** In graphite, only three of the valence electrons are used in bonding **[1]** to create a layer structure with mobile electrons between the layers, **[1]** which enable graphite to carry an electric current.

(c) In BF_3 there are 3 bonds and no lone pairs **[1]** hence the bond angle is 120°, **[1]** which places the F atoms as far apart as possible and minimises repulsions between them. **[1]** In NH_3 there are 3 bond pairs and one lone pair, resulting in a pyramidal molecule with an expected bond angle of somewhat less than 109° **[1]** (this being the bond angle between 4 bond pairs, e.g. in methane).

(d) Metals consist of close-packed arrangements of metal atoms which are held together by the mutual attraction of the nuclei and the mobile flux of electrons present. **[1]** This attractive force is non-directional and, given sufficient force, any metal atom can be displaced from its position in the lattice, taking the place of another without disrupting the lattice as a whole. **[1]** In sulphur the bonding is highly directional, forming S_8 rings, held together by van der Waals' forces. **[1]** The application of pressure

to the solid breaks these weak forces, creating only smaller crystals of the element: the smooth transition of one sulphur molecule from its original position to another is impossible. **[1]**

13 (a) An atom having the same atomic number but a different mass number/Two atoms of the same element having differing number of neutrons in their nuclei. **[2]**

(b) $^{79}_{35}Br$ $^{81}_{35}Br$ **[2]**

(c) 79.99 **[2]**

(d) **(i)**

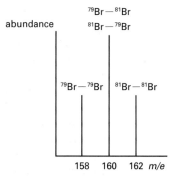

correct m/e **[3]**
1:2:1 ratio **[2]**

(ii) The origin of each peak is shown in the spectrum. **[1]** Notice that the three peaks are in a 1:2:1 **[1]** ratio because Br_2 has twice the probability of consisting of a mixture of ^{79}Br and ^{81}Br than it has of being composed entirely of just one of the isotopes. **[1]**

14 (a) **(i)** magnesium **(ii)** magnesium
(iii) barium **(iv)** barium **(v)** magnesium **[5]**

(b) **(i)** Atomic radius increases **[1]** because each successive atom contains one more shell of electrons **[1]** and the outermost electrons therefore experience a decreased nuclear attraction. **[1]**

(ii) The magnesium ion has the smaller radius **[1]** because it has the same electronic configuration as the sodium ion **[1]** but it has a higher effective nuclear charge, which attracts its electrons to a greater extent. **[1]**

(c) **(i)** Chlorine **[1]**

(ii) Graphite **[1]** because it has a high melting point and is inert (and it conducts electricity). **[1]**

(iii) Strontium is more reactive **[1]** than any (commercially viable) reducing agent. **[1]**

(d) (i) Strontium oxide **[1]** SrO **[1]**

(ii) $SrO(s) + H_2O(l) \rightarrow Sr(OH)_2(aq)$ **[2]**

15 (a) (i) chlorate(I), OS + 1 **[1]** chloride, OS − 1 **[1]**

(ii) It means that an element in a single species in the reaction has been both oxidized and reduced. **[2]**

(iii) The reaction is an equilibrium **[1]** so the addition of alkali would simply cause the equilibrium to shift to the right **[1]** which continuously alters the concentration of acid present as titration proceeds. **[1]**

(b) (i) ClO^- **[1]** because it can be oxidized and reduced **[1]** but Cl^- can be oxidized, but not further reduced. **[1]**

(ii) $3ClO^-(aq) \rightleftharpoons ClO_3^-(aq) + 2Cl^-(aq)$ **[2]**

(c) (i) Fluorine is a more powerful oxidizing agent than chlorine because the F—F bond is weaker than the Cl—Cl bond **[1]** and fluorine, being the smaller atom, accepts an electron more easily than chlorine. **[1]**

(ii) Fluorine atom is too small to accommodate (more than one) oxygen atoms around it **[1]**. Chlorine has vacant 3d-orbitals which can accept electron density from the O atoms (forming pi-bonds) **[1]** which give the Cl—O some double bond character/strengthens it. **[1]**

(iii) The H—F bond is very strong **[1]** and hydrogen bonds in liquid HF must also be broken. **[1]**

(d) (i) Bromine (some hydrogen bromide) and sulphur dioxide. **[3]**

(ii) The bromide ion is a more powerful reducing agent than the fluoride ion **[1]** because its outer electron is more easily lost **[1]** due to its larger size **[1]** and so reduces the sulphate ion.

16 (a) (i) Cs **[1]** $2Cs(s) + 2H_2O(l) \rightarrow$ $2CsOH(aq) + H_2(g)$ **[2]** It has the largest atom **[1]** so the outer electron is most readily lost. **[1]**

(ii) Mg **[1]** Solubility increases with increase in atomic number. **[1]**

(b) (i) Fe^{2+} **[1]** Loss of an electron creates a half-full 3d-orbital **[1]** which is particularly stable. **[1]**

(ii) Zn **[1]** Further ionization involves breaking into the 3d-orbital which is

energetically infeasible **[1]**. Also the +2 state gives the resulting ion (Zn^{2+}) a full 3d-orbital, which is relatively stable. **[1]**

(iii) MnO_4^- **[1]**, +7 **[1]**

(c) (i) I **[1]**, I_2 **[1]** The intermolecular forces are van der Waals' **[1]** the strength of which depends on the number of electrons in a molecule **[1]** and I_2 has the greatest number of electrons in its molecules. **[1]**

(ii) PbO **[1]** Lead has the largest atom in group 4 **[1]** so is most likely to form cations **[1]** and its lower (+2) oxide will be more basic than its higher oxide (+4). **[1]**

(iii) Any halogen **[1]** It reacts with water to form H^+ ions **[1]** because the energy needed to break the H—X bond is compensated for by the hydration enthalpies of the resultant ions **[1]** also the H—X bond is polarized in the sense $H^{\delta+}$—$X^{\delta-}$ **[1]** $HX(aq) + NaOH(aq) \rightarrow NaX(aq) +$ $H_2O(l)$ **[2]**

17 Missing rows are as follows:

(a)	tetrahedral	CH_4
(b)	pyramidal	NH_3
(c)	angular	H_2O
(d)	linear	HCl **[4 × 2]**

18 (a) $F_2(g)$ pale yellow, $Cl_2(g)$ pale green, $Br_2(l)$ deep red, $I_2(s)$ grey/black **[4]**

(b) (i) $KCl(s) + H_2SO_4(l) \rightarrow HCl(g) +$ $KHSO_4(s)$ **[2]**

(ii) Y acts as a proton donor (i.e. as a Brønsted–Lowry acid) **[1]** while water acts as a Brønsted–Lowry base **[1]** and accepts a proton: $HCl(aq) + H_2O(l) \rightarrow Cl^-(aq) +$ $H_3O^+(aq)$ **[1]**

(iii) Silver nitrate forms a precipitate of silver chloride when added to Z: $Ag^+(aq) + Cl^-(aq) \rightarrow AgCl(s)$. **[1]** However, in the presence of concentrated ammonia solution a complex is formed between the aqueous silver(I) ions in equilibrium with the precipitate of AgCl:

$Ag^+(aq) + 2NH_3(aq) \rightarrow Ag(NH_3)_2^+(aq)$, [1] which leads to the eventual disappearance of the precipitate or, if there is excess ammonia, the lack of any precipitate at all. [1] The reaction between ammonia and aqueous hydrochloric acid solution is exothermic [1] hence the second addition is carried out cautiously to prevent too great a temperature rise, which might drive out ammonia or hydrogen chloride fumes from the mixture. [1]

(c) $3ClO^-(aq) \rightleftharpoons 2Cl^-(aq) + ClO_3^-(aq)$. [1] The oxidation state of chlorine in ClO^- has both increased from $+1$ to $+5$ (in ClO_3^-) and decreased to -1 (in Cl^-). [1] This simultaneous oxidation and reduction of the same element in a single species characterises disproportionation. [1]

(d) (i) toxic to bacteria, soluble in water [2]

(ii) $Cl_2(aq) + SO_2(g) + 2H_2O(l) \rightleftharpoons 2Cl^-(aq) + SO_4^{2-}(aq) + 4H^+(aq)$ [3]

19 (a) (i) RAM of M = 39, so M is potassium [2]

(ii) $2M_2O(s) + O_2(g) \rightarrow 2M_2O_2(s)$ [2]

(iii) O_2^{2-} [1]

(iv) -1 [1]

(v) The -2 oxidation state of oxygen is the most stable, so reduction of the anion in X from OS -1 to OS -2 is likely, in which case the anion in X acts as an oxidizing agent. [2]

(b) (i) Radium would be softer than magnesium [1] because its atom is larger, leading to weaker metallic bonding due to the greater distance between the nucleus and the valence electrons. [1]

(ii) The Ra^{2+} ion would be larger than Ba^{2+} [1] because it has more electron shells and its electrons are better shielded from the nuclear charge. [1]

(iii) Radium metal would react more vigorously with water than calcium [1] because reaction with water involves the loss of the two s electrons, [1] which will be easier for radium because the atom is larger [1] and the electrons which are lost feel a lower nuclear attraction than those in the smaller calcium atom. [1]

(iv) Radium carbonate would not decompose into the corresponding oxide and carbon dioxide at a lower temperature than calcium carbonate [1] because the comparatively large radium ion [1] polarizes the carbonate ion to a lesser extent than calcium/the larger carbonate ion forms a relatively more stable lattice with the larger radium ion than does the smaller oxide ion. [1]

(v) Radium sulphate would not dissolve to a significant extent in water at room temperature [1] continuing the trend of decreasing solubility of the sulphates with increasing atomic number in the group. The increase in ionic radii reduces both the lattice enthalpies and the hydration enthalpies, [1] but reduces the hydration enthalpies to a greater extent, hence radium sulphate is insoluble because the lattice enthalpy cannot be compensated for by the hydration enthalpies of the resultant ions. [1]

20 (a) (i) hydrogen fluoride [1]

(ii) 'Weak' implies only partial/incomplete ionization in solution. [1]

(iii) The HF bond is particularly strong [1] and the hydrogen bonds between HF molecules must also be broken prior to ionization. [1]

(b) (i) hydrogen fluoride [1]

(ii) The HF bond is the most polar [1] due to the very high electronegativity of fluorine [1] and hence the intermolecular forces in HF are hydrogen bonds, which are considerably stronger than the intermolecular forces between molecules of the other hydrogen halides. [1]

(c) (i) van der Waals' forces [1]

(ii) HI because the iodine atom has more electrons than the bromine atom [1] so the van der Waals' forces are stronger in HI. [1]

(d) (i) sodium fluoride/chloride [1]

(ii) $NaX(s) + H_2SO_4(l) \rightarrow NaHSO_4(s) + HX(g)$ [2]

(e) (i) $2NaBr + 2H_2SO_4 \rightarrow Na_2SO_4 + SO_2 + 2H_2O + Br_2$

or

$8NaI + 5H_2SO_4 \rightarrow 4Na_2SO_4 + H_2S + 4H_2O + 4I_2$ [3]

(ii) The halide ion has been oxidized, **[1]** having increased in oxidation state from -1 to 0; the sulphate ion has been reduced, **[1]** having undergone a change in oxidation state from $+6$ to $+4$ (with bromide ion) or $+6$ to -2 (with iodide ion). **[1]**

(iii) Bromide and iodide ions are stronger reducing agents than fluoride or chloride **[1]** because they are larger ions **[1]** and the outer electron lost on oxidation is better shielded from the nuclear charge. **[1]**

21 (a) (i) V(III) because 0.17 V (the SO_4^{2-}/SO_3^{2-} potential) is less than 1.00 V and 0.34 V, **[1]** but greater than 0.26 V **[1]** which implies that sulphite can reduce V(V) to V(IV) and V(IV) to V(III), but not V(III) to V(II). **[1]**

(ii) $SO_3^{2-}(aq) + 2H^+(aq) + VO_2^+(aq) \rightleftharpoons$ $V^{3+}(aq) + SO_4^{2-}(aq) + H_2O(l)$ **[3]**

(iii) green **[1]**

(b) (i) Simultaneous oxidation and reduction of an element in a single species **[2]**

(ii) $2VO^{2+}(aq) \rightleftharpoons V^{3+}(aq) + VO_2^+(aq)$ **[2]** The $E^\ominus = -0.66$ so (because this is negative) **[1]** disproportionation will not occur under standard conditions. **[1]**

(iii) Changing the pH would have no effect on the tendency of V(IV) to disproportionate **[1]** because there are no hydrogen ions in the equation for the reaction. **[1]**

(c) (i) $+3$ **[1]**

(ii) dichlorotetraquovanadium(III) **[2]**

22 (a) (i) d-block element – last electron added to the atom entered a d-orbital **[1]** transition metal – they have several common characteristics, such as variable oxidation state, formation of coloured complex ions, the ability to act as catalysts and to form paramagnetic ions **[2]**

(ii) scandium and zinc **[2]**

(iii) Two from: variable oxidation state, formation of coloured complex ions, the ability to act as catalysts, formation of paramagnetic ions. **[2]**

(b) (i) $Cu = [Ar] 3d^{10} 4s^1$; $Cu^+ = [Ar] 3d^{10}$; $Cu^{2+} = [Ar] 3d^9$ **[3]**

(ii) Colourless **[1]** because the d-orbitals in

the Cu(I) ion are full **[1]** and there is therefore no possibility of d–d transitions taking place (which give transition metal ions their colour). **[1]**

(c) (i) $+6$ **[1]**

(ii) $n = 4$ **[1]**

(iii) $FeO_4^{2-}(aq) + 3I^-(aq) + 8H^+(aq) \rightleftharpoons$ $Fe^{3+}(aq) + 1\frac{1}{2}I_2(aq) + 4H_2O(l)$ **[3]**

(iv) tetrahedral **[1]**

(v) $[Ar] 3d^2$ **[1]**

23 (a) Copper(II) hydroxide, $Cu(OH)_2$ is insoluble (and is formed when the hydroxide ion present in the aqueous ammonia deprotonates the hexaquocopper(II) ion): **[1]** $Cu(H_2O)_6^{2+}(aq) + 2OH^-(aq) \rightarrow$ $Cu(OH)_2(H_2O)_4(s)$ [blue ppt.] $+ 2H_2O(l)$ **[2]** The $Cu(OH)_2(s)$ is in equilibrium with its constituent aqueous ions: $Cu(OH)_2(s) \rightleftharpoons Cu^{2+}(aq) + 2OH^-(aq)$ **[1]** and the addition of excess ammonia leads to the formation of the tetrammine copper(II) complex $Cu(NH_3)_4^{2+}$ **[1]** which displaces the above equilibrium to the right, **[1]** and eventually the copper(II) hydroxide dissolves completely. **[1]**

(b) Iron reacts exothermically with chlorine, forming iron(III) chloride **[1]** and with hydrogen chloride forming iron(II) chloride: **[1]**

$2Fe(s) + 3Cl_2(g) \rightarrow 2FeCl_3(s)$ **[1]** $Fe(s) + 2HCl(g) \rightarrow FeCl_2(s) + H_2(g)$ **[1]** Both chlorides are rapidly hydrolysed on contact with water to give aqueous solutions of the corresponding aquated cations: **[1]** $Fe^{2+}(aq)$ is pale green, **[1]** $Fe^{3+}(aq)$ is orange-brown. **[1]**

(c) In the case of zinc the reaction taking place in aqueous solution will be: $Fe^{2+}(aq) + Zn(s) \rightarrow Fe(s) + Zn^{2+}(aq)$ which has an $E^\ominus$ value of $(-0.44 + 0.76) = +0.32V$. **[1]** This implies that the zinc will corrode in preference to the iron. **[1]** However, since iron is a better reducing agent than tin (i.e. has a more negative $E^\ominus$ value) the reaction which occurs when tin is involved will be $Sn^{2+}(aq) + Fe(s) \rightarrow Sn(s) + Fe^{2+}(aq)$ which has an $E^\ominus$ value of

$(-0.14 + 0.44) = +0.30V$, **[1]** leading to corrosion of the iron. **[1]**

(d) Aqueous silver nitrate only reacts immediately with chloride ions, not the chlorine bonded covalently in the complex ion, hence A, B and C contain 1, 2 and 3 moles of chloride ions per mole. **[1]** A is $[CrCl_2(H_2O)_4]^+Cl^-.2H_2O$ **[1]** B is $[CrCl(H_2O)_5]^{2+}2Cl^-.H_2O$ **[1]** C is $[Cr(H_2O)_6]^{3+}3Cl^-$ **[1]**

Complex A can exhibit geometrical isomerism, having two structures, **[1]** one with the chlorine atoms *cis* to one another, the other with the chlorine atoms *trans* to one another. **[1]**

24 (a) The number of stable oxidation states shown by the d-block elements increases from scandium to manganese because there are more electrons available **[1]** which can be ionized or involved in covalent bonding. The ionization or promotional energies involved can be offset by the lattice enthalpies or bond enthalpies in the resulting compounds. **[1]** The ionization and promotional energies of the elements increase from Sc to Zn as a result of the increasing nuclear charge, **[1]** from which the d-electrons are only poorly shielded. **[1]** By the time iron is reached it seems that the energy required to involve all (eight) electrons in bonding cannot be recouped from compound formation and this continues to be the case thereafter. The number of known OS therefore declines after manganese. **[1]**

(b) Scandium and zinc belong in the d-block of the periodic table because the last electron to be added to their atoms entered the 3d orbital. **[1]** However, neither meets the criteria for inclusion in the transition elements because they both form only one OS, **[1]** which has no d-electrons in the case of Sc and full d-orbitals in the case of zinc. **[1]** Consequently, they have no coloured compounds. **[1]**

(c) Both manganese and iron have negative $E^\ominus$ values ($Fe^{2+}/Fe = -0.44V$, $Mn^{2+}/Mn = -1.18V$) **[1]** and are thus able to reduce hydrogen ions to hydrogen gas: **[1]** $M(s) + 2H^+(aq) \rightleftharpoons M^{2+}(aq) + H_2(g)$ **[1]** Manganese is a much more powerful

reducing agent than iron because the resulting Mn^{2+} ion has a particularly stable **[1]** d^5 electronic configuration, in which the d-orbital is half full. **[1]**

(d) Iron reacts with chlorine, forming iron(III) chloride **[1]** and with hydrogen chloride forming iron(II) chloride. **[1]** $2Fe(s) + 3Cl_2(g) \rightarrow 2FeCl_3(s)$; **[1]** $Fe(s) + 2HCl(g) \rightarrow FeCl_2(s) + H_2(g)$ **[1]** Iron(II) has a d^6 electronic configuration and readily loses an electron **[1]** to form iron(III) which has a particularly stable half-full 3d orbital. **[1]**

25 (a) (i) Zn $0 \rightarrow +2$, Mn $+4 \rightarrow +3$, NH_4^+ no change **[3]**

(ii) MnO_2 undergoes reduction as the cell reaction proceeds. **[1]**

(b) (i) $Mn^{3+}(aq) + e^- \rightarrow Mn^{2+}(aq)$ **[1]** $MnO_2(s) + 4H^+(aq) + e^- \rightarrow Mn^{3+}(aq) + 2H_2O(l)$ **[3]**

(ii) Simultaneous oxidation and reduction of the same element in a given species **[2]**

(iii) $2Mn^{3+}(aq) + 2H_2O(l) \rightleftharpoons MnO_2(s) + Mn^{2+}(aq) + 4H^+(aq)$ **[3]**

(iv) A decrease in pH (i.e. an increase in hydrogen ion concentration) decreases the feasibility of the disproportionation of $Mn^{3+}(aq)$. **[1]** The above equilibrium will be displaced towards the LHS **[1]** on the addition of more hydrogen ions. **[1]** (The first stage in the reverse reaction involves the protonation of $MnO_2(s)$.)

(c) (i) Mn^{3+} (Because it has the more positive $E^\ominus$ value.) **[1]**

(ii) Reduction of Mn^{3+} ($3d^4$) to Mn^{2+} ($3d^5$) is more favourable than the reduction of Fe^{3+} **[1]** because in the case of manganese this reduction results in the formation of a stable Mn^{2+} ion having a half-full 3d orbital, **[1]** while in the iron case Fe^{3+} already has the $3d^5$ configuration and reduction to Fe^{2+} is energetically unfavourable. **[1]**

26 (a) An element which, in at least one of its compounds, has a partially filled d orbital. **[2]**

(b) $Fe^{2+} = [Ar]\ 3d^6$, **[1]** $Fe^{3+} = [Ar]\ 3d^5$, **[1]** $Cu^+ = [Ar]\ 3d^{10}$, **[1]** $Cu^{2+} = [Ar]\ 3d^9$ **[1]**

(c) Three from: they form complexes, have more than one oxidation state, often act as catalysts, their ions may exhibit paramagnetism. **[3]**

(d) Iron: forms the ions Fe^{2+} and Fe^{3+}; is the catalyst in the Haber Process for ammonia manufacture; $Fe(CN)_6^{3-}$ is an example of an iron(III) complex ion; aqueous solutions of iron(III) salts are paramagnetic, compounds are coloured.

Copper: forms the ions Cu^+ and Cu^{2+}; examples of complexes include $Cu(NH_3)_4^{2+}$ and $CuCl_4^{3-}$ aqueous solutions of copper(II) salts are paramagnetic, compounds are coloured. **[3]**

27 (a) OS Mn in $MnO_4^- = +7$ and in $Mn^{2+} = +2$ **[2]**

(b) Reduction, because the oxidation state of manganese decreases from $+7$ to $+2$. **[2]**

28

(a)

		3d					4s
Cu$^+$	Ar	⇅	⇅	⇅	⇅	⇅	
Cr^{2+}	Ar	↑	↑	↑	↑		
Sc^{3+}	Ar						

(ii) Cu^+ and Sc^{3+} **[2]**

(iii) In Cu^+ the d orbitals are full, making electronic transitions between them impossible, **[1]** while in Sc^{3+} the d orbitals are empty. **[1]**

(b) (i) Co(III) **[1]** [Ar] $3d^6$ **[1]**

(ii) dichlorotetrammine cobalt(III) **[2]**

(iii) *dative* **[1]** covalent **[1]**

(iv) One mole of AgCl would be immediately precipitated **[1]** because only one mole of chlorine in the complex exists as free chloride ions, **[1]** the remainder being (dative) covalently bonded to the cobalt ion and Ag$^+$ reacts only with chloride ions. **[1]**

29 (a) Scandium **[1]** – it exhibits only one oxidation state, $+3$. **[1]**

Zinc **[1]** – the stable $+2$ oxidation state has a full d orbital. **[1]**

(b) (i) $+2$ **[1]**

(ii) (A) e.g. X$^-$, where X represents any halogen; **[1]** (B) e.g. H$_2$O or NH$_3$ **[1]**

(iii) square planar or tetrahedral **[1]**

(iv) *dative* **[1]** covalent **[1]**

(c) (i) Copper(II) hydroxide, $Cu(OH)_2$ is insoluble **[1]** and is formed when the hydroxide ion present in the aqueous ammonia deprotonates the hexaquocopper(II) ion: **[1]**

$Cu(H_2O)_6^{2+}(aq) + 2OH^-(aq) \rightarrow$
$Cu(OH)_2(H_2O)_4(s) + 2H_2O(l)$ **[2]**

(ii) The $Cu(OH)_2(s)$ is in equilibrium with its constituent aqueous ions:
$Cu(OH)_2(s) \rightleftharpoons Cu^{2+}(aq) + 2OH^-(aq)$ **[1]**
The addition of excess ammonia leads to the formation of the tetrammine copper(II) complex $Cu(NH_3)_4^{2+}$ **[1]** which displaces the above equilibrium to the right, **[1]** and eventually the copper(II) hydroxide dissolves completely. **[1]**

(iii) The precise colour observed depends on the nature of the ligand present in the complex **[1]** and upon its shape, **[1]** since both of these have an effect on the extent to which the d orbitals of the metal ion are split and hence the frequency of radiation absorbed when an electron is promoted. **[1]** In this example both changes occur: water is replaced by ammonia as the ligand and the resulting ammine complex is tetrahedral, not octahedral. **[1]**

30 (a) potentiometer/valve voltmeter **[1]**

(b) (i) Soak filter paper strip/string in saturated potassium nitrate solution. **[2]**

(ii) To allow ions to pass from one cell compartment to the other. **[1]**

(c) Standard electrode potential of the $M^{2+}/M = -0.28$ V **[2]**

(d) (i) $Cu^{2+}(aq) + M(s) \rightarrow Cu(s) + M^{2+}(aq)$ **[2]**
M is a better reducing agent than copper (it has a more negative $E^\ominus$) and so M can reduce copper(II) ions under standard conditions. **[2]**

(ii) $Cu(s) \mid Cu^{2+}(aq) \parallel M^{2+}(aq) \mid M(s)$ **[2]**

(iii) The cell e.m.f. would decrease. The overall cell e.m.f. is given by $E_R - E_L$, where E_R and E_L refer to the e.m.f.s of the RH and LH electrodes respectively, so diluting the solution of $Cu^{2+}(aq)$ reduces the value of E_R, leading to a decrease in the cell e.m.f. **[3]**

31 (a) formation of coloured **[1]** complexes **[1]**

(b) (i) Add aqueous nitric acid **[1]** followed by aqueous silver nitrate. **[1]** The formation of a white precipitate **[1]** which is soluble in aqueous ammonia **[1]** confirms the presence of free chloride ion.

(ii) Take equal masses of each isomer **[1]** and dissolve each in water. **[1]** Add excess acidified silver nitrate solution, **[1]** filter off the precipitate formed, **[1]** dry and weigh it. **[1]** The masses of silver chloride formed should be in the ratio 3:2:1 for **A**, **B** and **C** respectively. **[1]**

(c) octahedral **[1]**

(d) (i) $Cr^{3+} = [Ar] 3d^3$ **[1]**

(ii)

[2]

(iii) dative covalent **[1]**

(iv) ammonia **[1]**

32 (a) (i) From Sc to Mn the involvement of all the 3d and 4s electrons is possible, **[1]** the ionization energies increasing relatively slowly from Sc to Mn and being offset by hydration or bond energies in the resulting oxoanions. **[1]** The increasing nuclear charge makes the process less energetically favourable with increasing atomic number and at iron it becomes unfavourable to involve all three 3d and 4s electrons, leading to a maximum OS of 6, rather than the expected 8. **[1]** The highest oxidation state for cobalt might be +5, **[1]** falling even further short of theoretical maximum oxidation state than was the case with iron, owing to its even greater nuclear charge. **[1]**

(ii) No. **[1]** It has only a single oxidation state **[1]** and never forms coloured compounds or complex ions. **[1]**

(b) Mn^{3+} is unstable in aqueous solution, **[1]** being capable of oxidizing water to oxygen. **[1]** (It may, however, be kinetically stable, i.e. react slowly.) The gain of an electron by Mn^{3+} gives Mn^{2+}, which has a stable half-filled 3d orbital. **[1]** Fe^{3+} is stable in water, having a half-filled 3d orbital already and is therefore reluctant to gain the electron needed to reduce it to Fe^{2+}. **[1]**

(c) Initially a pale-blue precipitate of copper(II) hydroxide is formed as a result of deprotonation of the hexaquocopper(II) ion. **[1]** The addition of excess aqueous ammonia leads to complex formation between ammonia (the ligand) and aqueous copper(II) ions, **[1]** $Cu^{2+}(aq) + 4NH_3(aq) \rightarrow Cu(NH_3)_4^{2+}(aq)$ (the co-ordination number of copper is 4 in this complex ion). **[1]** This leads to the displacement of the following equilibrium to the right-hand side, eventually leading to complete dissolution on the precipitate of copper(II) hydroxide: **[1]** $Cu(OH)_2(s) \rightleftharpoons 2OH^-(aq) + Cu^{2+}(aq)$ **[1]**

(d) (i) +5 **[1]**

(ii) $BiO_3^-(aq) + 6H^+(aq) + 2e^- \rightleftharpoons Bi^{3+}(aq) + 3H_2O(l)$ **[2]**
$2Mn^{2+}(aq) + 5BiO_3^-(aq) + 14H^+(aq) \rightleftharpoons 2MnO_4^-(aq) + 5Bi^{3+}(aq) + 7H_2O(l)$ **[3]**

(iii) To force the equilibrium position to the RHS, **[1]** since H^+ ions appear on the LHS of the equation. **[1]**

(iv) 3.96% **[4]**

33 (a) (i)

	C	H	
%	88.89	11.11	
÷ RAM	7.41	11.11	**[1]**
÷ smaller	1	1.49	**[1]**
simplest ratio	2	3	**[1]**

EF is C_2H_3. As parent ion peak is at 54, molecular formula is C_4H_6.

(ii) Possible isomers are the diene $CH_2=CH-CH=CH_2$ and the cyclic alkene
H_2C-CH
$|\quad\quad||$
H_2C-CH **[2]**

(b) (i) Since 20 cm³ of X reacts with twice this volume of hydrogen, **[1]** X must contain two C=C bonds. **[1]** The diene $CH_2=CH-CH=CH_2$ or must be the correct structure.

(ii) Buta-1,3-diene **[1]**

(c) m/e 54 corresponds to the parent ion peak (M) due to $CH_2=CH-CH=CH_2^+$ **[1]** m/e 40 corresponds to the fragment $CH_2=CH-CH^+$ (M − 14) resulting from the loss of CH_2 from the parent ion **[1]** m/e 27 corresponds to the fragment $CH_2=CH^+$ resulting from the breaking of the C—C bond **[1]**

00

Practice in Chemistry

34 (a) (i)

bonds broken (kJ/mol)	bonds formed (kJ/mol)
C—H +412	C—Cl −338
Cl—Cl +242	H—Cl −431
total +654 **[1]**	−769 **[1]**

Enthalpy change = +654 + (−769) **[1]**
= −115 kJ mol⁻¹ **[1]**

(ii) The reaction is exothermic (i.e. thermodynamically feasible) **[1]** so its reluctance to occur in the dark must relate to a kinetic effect. The light provides the activation energy for the reaction. **[1]**

(b) $CH_4(g) + Cl\cdot(g) \rightarrow CH_3\cdot(g) + HCl(g)$ **[2]**
$CH_3\cdot(g) + Cl_2(g) \rightarrow CH_3Cl(g) + Cl\cdot(g)$ **[2]**

(c) (i) The suggested (incorrect) propagation steps would lead to the formation of molecular hydrogen **[1]** as a result of recombination of the H free radicals **[1]** (no such hydrogen is found). Additionally, the detection of traces of ethane in the products strongly supports the involvement of CH₃ free radicals in the mechanism. **[1]**

(ii) The answer cannot be energetic, since the reactants and products are identical for both sets of equations, **[1]** so the enthalpy changes must be the same. **[1]** Some sort of kinetic effect must be involved, **[1]** perhaps the decreased likelihood/effectiveness of collision between H and Cl₂ compared to that between CH₃ and Cl₂. **[1]**

35 (a) (i)

pentane **[1]**
2-methylbutane **[1]**
2,2-dimethylpropane **[1]**

(ii) The intermolecular forces are van der Waals' and due largely to the electrons on the carbon atoms. **[1]** In the unbranched chain the carbon atoms are able to approach each other more closely than is possible in 2-methylbutane, **[1]** which in turn can approach each other more closely than is possible in 2,2-dimethylpropane. **[1]** Hence the boiling point order (highest first) is: pentane > 2-methylbutane > 2,2-dimethylpropane. **[1]**

(b) (i)

are chiral

correct mirror image **[1]**

(ii) Four different groups bonded to one carbon atom **[1]** which leads to the molecule having a non-superimposable mirror image. **[1]**

(c) (i) $C_xH_y(g) + (x + y/4)O_2 \rightarrow x\,CO_2(g) + y/2\,H_2O(l)$ **[3]**

(ii) $x = 4$ **[2]**

(iii) $y = 8$ **[3]**

(d)

trans-but-2-ene **[1]**
cis-but-2-ene **[1]**

Rotation about the C=C double bond is impossible, so the two isomers are distinct. **[1]**

36 (a) (i) *Homolytic* implies that the bonding pair of electrons is shared equally between the resulting species (i.e. free radicals, as opposed to ions, result). **[2]**
A *free radical* is a species having an unpaired electron. **[1]**

198

(ii) Sunlight provides the energy necessary to break the Br—Br bond, thus generating the free radicals needed to initiate the reaction. **[1]**

(b) (i) Initiation: $Br_2(g) + hf \rightarrow 2Br\cdot(g)$ **[2]**

Propagation:

$CH_4(g) + Br\cdot(g) \rightarrow HBr(g) + CH_3\cdot(g)$ **[2]**

$CH_3\cdot(g) + Br_2(g) \rightarrow CH_3Br(g) + Br\cdot(g)$ **[2]**

(ii) The initial monobrominated product can react further to form species containing more than one bromine atom: **[1]** a mixture of products is formed and their separation is tedious. **[1]**

(c) (i) The peaks at m/e 94 and 96 correspond to $CH_3{}^{79}Br$ and $CH_3{}^{81}Br$, respectively. **[1]** They are of approximately equal intensity because the relative abundances of ^{79}Br and ^{81}Br are about the same. **[1]**

(ii) There must be two bromine atoms per molecule. **[1]** First because the RMM is in the correct range for CH_2Br_2, **[1]** second because the three peaks in the m/e range 172–176 with relative intensities 1:2:1 is consistent with this. **[1]**

m/e 172 corresponds to $CH_2{}^{79}Br^{79}Br$

m/e 174 corresponds to $CH_2{}^{79}Br^{81}Br$ (or $CH_2{}^{81}Br^{79}Br$, which gives this peak twice the intensity of those at m/e 172 and m/e 176)

m/e 176 corresponds to $CH_2{}^{81}Br^{81}Br$. **[1]**

(iii) The peak in the mass spectrum at m/e 30 may well be due to $C_2H_6{}^+$ from ethane **[1]** formed by the termination reaction between two methyl radicals: $2CH_3\cdot(g) \rightarrow C_2H_6(g)$. **[1]** The formation of ethane supports the mechanism because methyl free radicals must have been present as intermediates. **[1]** An alternative propagation mechanism, such as:

$CH_4(g) + Br(g) \rightarrow H\cdot(g) + CH_3Br(g)$

then $H\cdot(g) + Br_2(g) \rightarrow HBr(g) + Br\cdot(g)$

would lead to traces of hydrogen in the products, but no higher alkanes. **[1]**

37 (a) The law of conservation of energy/Hess's law. **[1]**

(b) Standard enthalpy change of formation of propane **[2]**

(c) $+218$ kJ mol^{-1} **[3]**

(d) $+3996$ kJ mol^{-1} **[2]**

(e) (i) Correct energy cycle. **[1]**

$\Delta H^{\ominus}_{atm}(C_3H_6) = 3436$ kJ mol^{-1} **[5]**

(ii) 612 kJ mol^{-1} **[3]**

38 (a) (i) (bromobenzene) + magnesium **[1]** in *dry* ethoxyethane **[1]**

(ii) $C_6H_5Br + Mg \rightarrow C_6H_5MgBr$ **[1]**

(iii) No naked flames **[1]** because ethoxyethane is highly flammable. **[1]**

methanal   ethanal

propanone   epoxyethane   **[4]**

(b) (i)

(ii) B is a methyl secondary alcohol **[1]** and therefore gives a positive triiodomethane (iodoform) test. **[1]** To iodine in potassium iodide solution, **[1]** add aqueous sodium hydroxide dropwise until the mixture is pale yellow. **[1]** Then add a few drops of B and warm. The appearance of a pale yellow precipitate having an antiseptic smell is the positive test result. **[1]**

(iii) B, having four *different* groups bonded to a carbon atom is chiral. **[1]** This implies that it is able to rotate the plane of plane polarized light, **[1]** which can be tested by placing it in a polarimeter. A would have no effect on the plane of plane polarized light. **[1]**

39 (a) (i) A positively charged species which is attracted to/tends to react with electron rich sites in a molecule. **[1]**

(ii) $NO_2{}^+$ **[1]**

(iii) $HNO_3 + H_2SO_4 \rightarrow H_2NO_3{}^+ + HSO_4{}^-$

$H_2NO_3{}^+ \rightarrow H_2O + NO_2{}^+$ **[2]**

(iv)

[3]

(b) (i) $AlCl_3 + Cl_2 \rightarrow [AlCl_4]^- \, Cl^+$ **[2]**

(ii) ethanoyl chloride **[1]**

CH_3COCl **[1]**

(c) (i) benzoic acid **[1]**

(ii) $KMnO_4$ **[1]**

(iii) $MnO_4^- + 8H^+ + 5e^- \rightarrow Mn^{2+} + 4H_2O$ **[2]**

(iv) The oxidation number of manganese has decreased from $+7$ to $+2$ and this corresponds to reduction. **[2]**

40 (a) (i) Step 1: concentrated **[1]** nitric + sulphuric acids **[1]** temperature kept below 30 °C **[1]**

Step 2: tin metal **[1]** concentrated hydrochloric acid **[1]** heat under reflux **[1]**

[1]   **[1]**   **[1]**

[1]   **[1]**   **[1]**

(b) (i)

[2]

(ii) The ionic species which results on acidification benefits from considerable hydration enthalpy in water and this allows the hydrogen bonds in the (unprotonated) phenylamine to be overcome and it dissolves. Under neutral conditions, although phenylamine itself is hydrogen bonded, these are not as strong as the hydrogen bonds in water and the energy required to disrupt the hydrogen bonds in water cannot be recouped by formation of hydrogen bonds of comparable strength between the water and the phenylamine, hence it fails to dissolve significantly. **[4]**

(iii)

[2]

(iv) $C_6H_5N_2^+Cl^-(aq) + H_2O(l) \rightarrow$
$C_6H_5OH(aq) + HCl(aq) + N_2(g)$ **[3]**

(v) diazotization **[1]**

(vi)

Such precipitates are usually red or orange. **[3]**

41 (a) (i) An increase in temperature increases the rate **[1]** principally as a result of the increase in the average energy of the reactant molecules, **[1]** leading to a greater number having energies greater than the activation energy **[1]** and hence being able to react on collision. **[1]** A second, but minor effect, is the increase in the collision rate **[1]** as a result of the greater speed of the reactant molecules **[1]**.

(ii) A catalyst allows the activation energy to be reduced **[1]** by letting the reaction proceed by a different reaction pathway. **[1]** At a given temperature more reactant molecules will have energies in excess of the (now lower) activation energy **[1]** and will therefore be able to react when they collide. **[1]**

(iii) Increasing the concentration brings the reactants into closer proximity with each other. **[1]** This increases the probability that the reactant molecules will collide with one another. **[1]** The reactants must collide before they can react **[1]** (but they will not if their combined energy does not exceed the activation energy for the reaction).

(b) (i) Using the first two left-hand entries in the table, the order w.r.t. A = 1 (first order) since doubling its concentration from 1.0 to 2.0 $mol\ dm^{-3}$ while keeping the concentration of B constant leads to a doubling (i.e. 2^1) increase in the rate of the reaction. **[2]**

(ii) Using the first two right-hand entries in the table, the order w.r.t. B = 2 (second order) since doubling its concentration from 1.0 to 2.0 $mol\ dm^{-3}$ while keeping the concentration of A constant leads to a four-fold (i.e. 2^2) increase in the rate of the reaction. **[2]**

(iii) rate $= k[A][B]^2$ **[2]**

(iv) Overall order $= 2 + 1 = 3$ **[1]**

(v) 1.2×10^{-2} **[2]**
the units of $k = mol^{-2}\ dm^6\ s^{-1}$ **[2]**

42 (a) (i)

[2]

(ii) It is able to rotate the plane of polarization **[1]** of plane-polarized light. **[1]**

(b) (i) free-radical/homolytic substitution **[1]**

(ii) initiation:

$Cl_2(g) + hf \rightarrow 2Cl\cdot(g)$ **[2]**

propagation:

$C_2H_6(g) + Cl\cdot(g) \rightarrow HCl(g) + C_2H_5(g)$
$C_2H_5(g) + Cl_2(g) \rightarrow C_2H_5Cl(g) + Cl\cdot(g)$

[4]

termination:

$2C_2H_5(g) \rightarrow C_4H_{10}(g)$

or

$C_2H_5(g) + Cl(g) \rightarrow C_2H_5Cl(g)$ **[2]**

43 (a) (i) bromomethane/chloromethane

CH_3Br/CH_3Cl **[2]**

(ii) CH_3^+ **[1]**

(iii) $CH_3Cl + AlCl_3 \rightarrow CH_3^+[AlCl_4]^-$ **[2]**

(iv)

$+ CH_3Cl$

[1] **[1]** **[1]** **[1]**

(b) Benzene fails to undergo addition reactions with bromine. **[1]**
Both benzene and ethene can be hydrogenated using hydrogen gas and a nickel/platinum catalyst. **[1]**

(c)

[3]

44 (a) (i) The sum of the powers to which each reactant is raised in the rate equation for the reaction. **[2]**

(ii) The proportionality constant, k, in the rate equation: rate $= k[A][B]$ (it is the reaction rate when all reactants have unit concentrations). **[2]**

(b) (i) 1st order w.r.t hydrogen ions **[1]**
1st order w.r.t. methyl ethanoate **[1]**

(ii) 2nd order overall **[1]**

(iii) rate $= k[H^+][CH_3COOCH_3]$ **[1]**

(iv) $0.002 \ mol^{-1} \ dm^3 \ min^{-1}$ **[3]**

(c) (i) The minimum (combined) energy **[1]** which two colliding particles must have in order to react. **[1]**

(ii) A catalyst provides an alternative route of lower activation energy, **[1]** so at both temperatures more of the reactants have energies greater than the activation energy and the reaction rate increases with a catalyst present **[1]**.

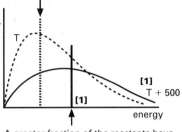

A greater fraction of the reactants have energies in excess of the activation energy at the higher temperature (T + 500) than that at the lower temperature (T) **[1]**, so the rate is faster at the higher temperature. **[1]**
correct shapes **[1]**

(iii) See graph for answers. **[4]**

45 (a) (i)

[1] **[1]**

$Br \!-\! Br$ $\qquad$ $:Br^-$

$CH_2 = CH_2 \longrightarrow CH_2 - CH_2Br \longrightarrow CH_2Br - CH_2Br$
$\qquad\qquad\qquad\qquad +$

[1] **[1]**

(ii) The π bond is a region of high electron density, capable of furnishing the electron pair required to bond the first bromine atom: **[1]** this could not happen in an alkane without complete breaking of the C—C bond. **[1]** Of at least equal importance is the ability of the π electrons to be polarized by the incoming bromine which leads to the drift of electron density towards one of the carbon atoms, forming the initial C—Br bond. **[1]**

(iii) electrophilic **[1]** addition **[1]**

(b) (i) The intermediate carbocation ($^+CH_2—CH_2Br$ in the above mechanism) **[1]** is equally susceptible to attack by Cl^- (furnished by the added sodium chloride) **[1]** as it is to attack by the Br^- formed by heterolytic fission of the Br—Br bond, hence both $C_2H_4Br_2$ and C_2H_4BrCl are formed. **[1]**

(ii) The formation of $C_2H_4Cl_2$ requires the formation of a carbocation containing chlorine (i.e. $^+CH_2—CH_2Cl$ in this case) **[1]** which can only result from the reaction between ethene and *molecular* chlorine. **[1]** Since no molecular chlorine is present the chlorine containing product is limited to that formed when the chloride ions attack the carbocation intermediate. **[1]**

(c) (i) $CH_3-CH_2-CH_2Br$ and $CH_3-CHBr-CH_3$ **[2]**

(ii)

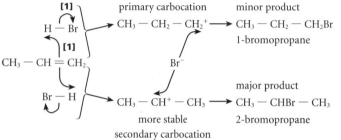

less stable primary carbocation — minor product

$CH_3 - CH_2 - CH_2^+ \longrightarrow CH_3 - CH_2 - CH_2Br$

1-bromopropane

major product

$CH_3 - CH^+ - CH_3 \longrightarrow CH_3 - CHBr - CH_3$

more stable secondary carbocation — 2-bromopropane

Hydrogen bromide can add across the double bond in an unsymmetrical alkene in two different ways – in the case of propene to give 1-bromo or 2-bromopropane. The mechanism is shown above. The positive charge in the carbocation intermediate is stabilized by the electron-releasing effect of alkyl groups **[1]** and since there are two bonded to the carbon carrying the positive charge in the secondary carbocation, **[1]** it is more stable than the corresponding primary carbocation. **[1]** There is another aspect to be considered (not often easy to quantify) and that concerns the relative *rates* of formation of the two carbocations: if the secondary one is formed *faster* than the primary one, then 2-bromopropane will result for purely kinetic reasons and not solely because of any difference in thermodynamic stability between the primary and secondary carbocations. **[1]**

46 (a) (i)

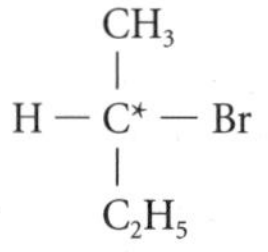

$$\begin{array}{c} CH_3 \\ | \\ H - C^* - Br \\ | \\ C_2H_5 \end{array}$$ **[1]**

(ii) A molecule which can rotate the plane of plane-polarized light/has a non-superimposable mirror image. **[2]**

(iii) Four different groups bonded to the same carbon atom (C^* in the formula). **[1]**

(b) (i)

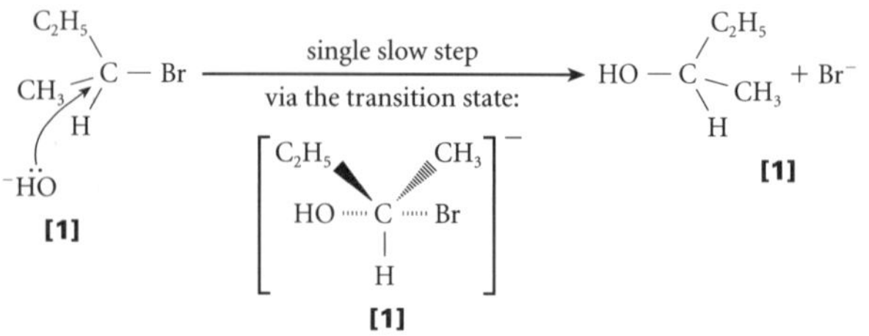

$$\begin{array}{c} C_2H_5 \\ \diagdown \\ CH_3 \diagup C - Br \\ | \\ H \end{array} \xrightarrow[\text{via the transition state:}]{\text{single slow step}} \begin{array}{c} C_2H_5 \\ \diagdown \\ HO - C \diagdown CH_3 + Br^- \\ | \\ H \end{array}$$ **[1]**

^-HO **[1]**

$$\left[\begin{array}{c} C_2H_5 \quad CH_3 \\ \diagdown \quad \diagup \\ HO \cdots C \cdots Br \\ | \\ H \end{array} \right]^-$$ **[1]**

(ii) In the S_N2 mechanism the starting bromoalkane can be considered to be turned 'inside out', rather like an umbrella in a strong wind **[1]** and at no point during the mechanism is there an intermediate species which can be attacked by the incoming nucleophile other than from the side opposite the departing bromide ion. **[1]** This ensures that a chiral alcohol will result from the pure (+)- or (−)-form of a chiral bromoalkane. **[1]**

(iii) At 80 °C the S_N1 mechanism becomes more likely **[1]** because more energy is available to break the C—Br bond in the slow rate-determining step. **[1]** The planar carbocation formed can be attacked equally from both sides. **[1]**

(c) (i) The groups in 2-bromo-2-phenylbutane experience considerable steric crowding **[1]** and the formation of the intermediate carbocation relieves this crowding as the bond angle increases from 109° to 120°. **[1]** The intermediate carbocation in the S_N1 mechanism is stabilized by the electron-releasing effect of the alkyl and phenyl groups. **[1]** The S_N2 mechanism is also unlikely in 2-bromo-2-phenylbutane because the bulky groups surrounding the carbon to which the bromine is bonded inhibit attack by the nucleophile. **[1]**

(ii) The intermediate carbocation is planar and hence equally likely to be attacked by a hydroxide ion from either side, **[1]** leading to the formation of equal amounts of the (+)- or (−)-forms of the alcohol. **[1]** The resulting reaction mixture shows no net optical activity because the tendency of one optical isomer to rotate the plane of the light in one sense, is exactly balanced by the rotation of the light in the opposite sense by the other isomer. **[1]** (It is a useful exercise to draw the resulting alcohols for yourself and to verify that they are a non-superimposable mirror image pair.)

(d)

romobutane →

$$\underset{H_3C}{\overset{H}{>}}C=C\underset{H}{\overset{CH_3}{<}} \text{[1]} \qquad \underset{H_3C}{\overset{H}{>}}C=C\underset{CH_3}{\overset{H}{<}} \text{[1]} \qquad \underset{C_2H_5}{\overset{H}{>}}C=C\underset{H}{\overset{H}{<}} \text{[1]}$$

trans-but-2-ene   *cis*-but-2-ene   but-1-ene
└────── **[1]** ──────┘   **[1]**

$$\underset{H_3C}{\overset{C_6H_5}{>}}C=C\underset{H}{\overset{CH_3}{<}} \text{[1]} \qquad \underset{H_3C}{\overset{C_6H_5}{>}}C=C\underset{CH_3}{\overset{H}{<}} \text{[1]} \qquad \underset{C_2H_5}{\overset{C_6H_5}{>}}C=C\underset{H}{\overset{H}{<}} \text{[1]}$$

2-bromo-
-phenylbutane →

trans-2-phenylbut-2-ene *cis*-2-phenylbut-2-ene 2-phenylbut-1-ene
└────── **[1]** ──────┘   **[1]**

47 **(a)** 0.16 g **[2]**, **(b)** 47.8 g **[2]**, **(c)** 4.69 million tonnes **[2]**, **(d)** 0.75 g **[3]**, **(e)** 2.70 g of cobalt(II) carbonate, 4.30 g of ethanedioic acid-2-water, 6.27 g of potassium ethanedioate-1-water and 2.72 g of lead(IV) oxide. **[6]**

48 **(a)** **(i)** 71.8% **[2]**, **(ii)** 78.5% **[2]**, **(iii)** 56.3% **[1]**, **(b)** 60.0%. **[3]**

49 **(a)** $HCl + NaOH \rightarrow NaCl + H_2O$ **[1]**, concentration = 0.090 M **[2]**

(b) Equation as for **(a)**, concentration = 0.057 M **[2]**

(c) $H_2SO_4 + K_2CO_3 \rightarrow K_2SO_4 + H_2O + CO_2$ **[1]**, concentration = 0.26 M **[2]**

(d) $Ba(OH)_2 + 2HCl \rightarrow BaCl_2 + 2H_2O$ **[1]**, mass dissolved in 1 dm³ = 26.7 g **[3]**

(e) $H_2C_2O_4 + 2NaOH \rightarrow Na_2C_2O_4 + 2H_2O$ **[1]**, concentration = 0.053 M **[2]**

(f) $H_3AsO_4 + 3NaOH \rightarrow Na_3AsO_4 + 3H_2O$ **[1]**, concentration **(i)** = 0.047 mol dm^{-3}, **(ii)** = 6.72 g dm^{-3} **[3]**

(g) Equation as for **(a)**, volume = 20.8 cm³ **[2]**

(h) $2NH_3 + H_2SO_4 \rightarrow (NH_4)_2SO_4$ **[1]**, volume = 14.3 cm³ **[2]**.

50 **(a)** 0.030 mol dm^{-3} **[2]**, **(b)** 0.00030 mol **[1]**, **(c)** 0.00060 mol **[1]**, **(d)** 0.015 mol dm^{-3} **[1]**.

51 No. of moles of acid used = 0.020 = no. of moles of alkali in excess **[1]**, no. of moles of alkali used initially = 0.040 **[1]**, therefore no. of moles of alkali reacted with the ammonium chloride present = (0.040 − 0.020) = 0.020 **[1]**, therefore no. of moles of ammonium chloride present = 0.010 **[1]**, therefore no. of grams of

ammonium chloride present = 0.010 × 53.5 = 0.535 g **[1]**, therefore percentage purity = 97.3%. **[1]**

52 Molar mass of sulphamic acid = 97 g mol^{-1} **[1]**, therefore no. of moles of sulphamic acid used = 1.50 g/97 g mol^{-1} = 0.0155 mol **[1]**, therefore no. of moles of sulphamic acid in 20 cm³ = 20 cm³/250 cm³ × 0.0155 mol = 0.00124 mol **[1]**, no. of moles of sodium carbonate used = 0.00062 mol **[1]**, therefore no. of moles of hydrogen ions present = 0.00124 **[1]**, therefore one mole of sulphamic acid has one mole of replaceable hydrogen ions. **[1]**

53 No. of moles of acid used = 0.0071 **[1]**, therefore no. of moles of sodium hydroxide left in 20 cm³ of solution = 0.0071 **[1]**, original no. of moles of sodium hydroxide used = 0.02 **[1]**, therefore no. of moles of sodium hydroxide reacted with the salt = (0.02 − 0.0071) = 0.0129 **[1]**, therefore no. of moles of ammonium ions present in 20 cm³ of solution = 0.0129 **[1]**, therefore no. of moles of microcosmic salt present in 20 cm³ of solution = 0.0129 **[1]**, therefore no. of moles of microcosmic salt originally used = 0.0645 **[1]**, therefore molar mass of microcosmic salt = 10.0 g/0.0645 g mol^{-1} = 155 g mol^{-1} **[1]**, therefore x = 1. **[1]**

54 **(a)** 0.025 mol **[1]**, **(b)** $2NaOH + H_2SO_4 \rightarrow H_2SO_4 + 2H_2O$ **[1]**, **(c)** 0.0008 mol of NaOH in a pipette volume **[1]**, therefore 0.008 mol in the standard flask **[1]**

(d) 0.004 mol **[1]**, **(e)** 0.072 g **[1]**, **(f)** phenolphthalein, methyl orange etc. **[1]**.

55 Propene has relatively small molecules with weak forces of attraction between them, therefore low melting and boiling points. Poly(propene) is a polymer made by joining together many small molecules into a long chain; therefore much larger forces between much bigger molecules, therefore much higher melting point and also gives rise to strength so capable of many important applications. **[4]**

56 **(a)** Standard enthalpy change of reaction = +96.1 kJ mol^{-1} **[2]**

(b) Standard enthalpy change of reaction = +91.6 kJ mol^{-1} **[1]**

(c) Ionic equation is:

$$2HCO_3^-(s) \rightarrow CO_3^{2-}(s) + CO_2(g) + H_2O(l)$$

The values obtained are closely similar because it is essentially the same reaction that is occurring in each case. **[2]**

57 Standard enthalpy change of reaction
$= -3.5$ kJ mol^{-1}. **[2]** The value is unusually small for an enthalpy change of reaction – most reactions are markedly exothermic or endothermic. **[1]**

58 (a) (i) The addition of sulphate ions moves the first equation to the left **[1]**, therefore solution turns blue. **[1]**

(ii) The blue-green solution had reached a state of equilibrium. (A) is warmed – first equilibrium moves to the left, (B) is cooled – first equilibrium moves to the right, (C) remained unchanged. **[3]**

(b) Forward reaction is endothermic – as it is favoured by lowering the temperature (or, raising the temperature favours the backward reaction). **[1]**

(c) Adding a lot of chloride ions moves the second equilibrium to the right **[1]** forming yellow $[CuCl_4]^{2-}$(aq) ions (the solution may well, in practice turn green – a mixture of blue and yellow). **[1]**
Diluting the solution will cause the reaction to move back to the left as a result of the extra water added, **[1]** So the mixture will turn back to blue (though, by now, very pale). **[1]**

59 Absorption of water will move the equilibrium towards the left **[1]**, so reforming the pink $[Co(H_2O)_6]^{2+}$(aq) ion. **[1]** On heating water is lost. Loss of a product will cause the equilibrium to move towards the right **[1]**, therefore becoming blue as the $[CoCl_4]^{2-}$(aq) ion is reformed. **[1]**

60 (a) (Iodine gives the yellowish-brown colour in the aqueous layer and the purple colour in the organic layer.) The observation shows that ammonia reacts with the iodine, forming a colourless product. **[1]**

(b) Organic layer – iodine removed by ammonia so colour is lost **[1]**. Aqueous layer – iodine colour also removed by added ammonia. **[1]**

(c) Both the reaction in the organic layer and the reaction in the aqueous layer are reversible **[1]** – the addition of acid to the organic layer reverses

the effect of the alkaline ammonia solution, **[1]** regenerating the iodine, thus giving the golden yellow colour in the water layer and the purple colour in the organic layer. **[1]**

61 (a) $^{37}_{17}Cl$, **(b)** 17, **(c)** 37, **(d)** 28, **(e)** 30, **(f)** 58, **(g)** $^{65}_{29}Cu$, **(h)** 29, **(i)** 36, **(j)** $^{74}_{34}Se$, **(k)** 34, **(l)** 74, **(m)** $^{81}_{35}Br$, **(n)** 35, **(o)** 46. **[15]**

62 (a) Should include the following main points: vapour of substance bombarded by electrons, **[1]** positive ions formed, **[1]** passed through slit to form fine beam, **[1]** accelerated by electric field, **[1]** deflected by magnetic field, **[1]** deflection depends on m/e of ions, **[1]** focused onto a detector. **[1]**

(b)

molecule	number of peaks expected in parent ion region
BeCl$_2$	three – both 35, both 37, one 35 + one 37
PCl$_3$	four – all 35, two 35 + one 37, one 35 + two 37, all 37
HCN	four – 12 + 14, 12 + 15, 13 + 14, 13 + 15
ClO$_2$	six – 35 + two 16, 35 + one 16 and one 17, 35 + two 17, 37 + two 16, 37 + one 16 and one 17, 37 + two 17 **[4 × 2]**

63 (a) B, **(b)** C, **(c)** D, **(d)** E, **(e)** A. **[5 × 1]**

64 (a) (i) NaCl is an ionic solid, the inter-ionic forces are strong and many have to be broken per mole to melt the solid. **[1]** In HCl there will be van der Waals' forces and dipole–dipole forces, **[1]** which are much weaker than the inter-ionic forces in NaCl and so HCl will boil at a much lower temperature. **[1]**

(ii) H$_2$O has strong intermolecular hydrogen bonds **[1]** as a result of the large electronegativity difference between O and H. **[1]** In CH$_4$ there are only weak van der Waals' forces and so CH$_4$ boils at a lower temperature. **[1]**

(iii) The intermolecular bonds in both I$_2$ and F$_2$ are exclusively van der Waals', **[1]** but iodine contains more electrons than does fluorine **[1]** and since the strength of such intermolecular forces depends on the number of electrons, the boiling point of iodine exceeds that of fluorine. **[1]**

(b) (i) Pentane is a straight chain alkane which increases the strength of van der Waals' forces because a larger proportion of the carbon atoms in one molecule can be in close proximity to those in an adjacent molecule. **[1]** In 2,2-dimethylpropane the branched chain prevents close contact between some of the carbon atoms in adjacent molecules **[1]** and reduces their contribution to the intermolecular forces. **[1]**

(ii) Butan-1-ol has an O—H group and hence intermolecular hydrogen bonding is possible, **[1]** which is much stronger than the van der Waals' and dipole–dipole forces in ethoxyethane, **[1]** which lacks an O—H group. **[1]**

(iii) The strength of metallic bonding depends on the number of available valence electrons which can contribute to the mobile 'sea' of electrons. **[1]** Potassium has one electron outside the argon core and hence metallic bonding is relatively weak **[1]** compared to that in copper which is able to use both its single 4s electron and, to some extent, the electrons in its 3d shell. **[1]**

(iv) Methane is a symmetrical molecule with only slightly polar C—H **[1]** bonds – the intermolecular forces are very weak indeed and its boiling point is very low in consequence. **[1]** By contrast, ammonia is extensively hydrogen bonded as a result of the polar N—H bonds and so has a much higher boiling point. **[1]**

65 (a) Simple molecular – e.g. water, methane, ammonia
giant molecular – e.g. silicon(IV) oxide, diamond
ionic – e.g. sodium chloride, lithium nitrate, aluminium fluoride. **[3]**

(b) (i) Each atom in the structure is linked to every other atom *via* covalent bonds, **[1]** giving a structure which requires many (strong) bonds to be broken to distort it **[1]** (i.e. to move one atom with respect to the rest), so such substances are often hard. **[1]**

(ii) Simple molecular substances contain no ions (by definition) **[1]** and their electrons are not free to move, **[1]** being localized in their covalent bonds, **[1]** hence they are unable to conduct electricity.

(iii) Ionic substances consist of regular arrays of cations and anions arranged such that ions of like charge are not next to each other in the structure. **[1]** A sharp blow can slide one layer of ions with respect to its neighbours, **[1]** thus bringing ions of like charge into contact. The repulsive forces between these like charges cause the crystal to shatter. **[1]**

66 (i) Xenon has more electrons than argon and, because the only possible inter-atomic forces are van der Waals' **[1]** and these depend on the number of electrons present, the boiling point of xenon is higher than that of argon. **[1]**

(ii) The H—F bond in hydrogen fluoride is considerably more polar than the H—Cl bond **[1]** and HF contains intermolecular hydrogen bonds which are much stronger than the corresponding van der Waals' and dipole–dipole forces in HCl. **[1]** Consequently at 15 °C the intermolecular forces in HF are sufficiently strong to allow it to exist as a liquid, while those in HCl are too weak and the substance boils. Hence hydrogen chloride is gaseous at this temperature. **[1]**

67 (a) propanal **[1]**

(b) (i) butanone **[1]**

(ii) Because the two and three carbon atoms are the same, it just depends which end you start to count from. Convention states take the lower, but to include 2 is a tautology as it is not required. **[1]**

(c) 1-hydroxybutan-3-one **[1]**

(d) 1-bromobutan-3-one **[1]**

(e) 2-chloro-1-hydroxybutan-3-one **[1]**

(f) 2-hydroxy-2-methylpropanal **[1]**

(g) (i) CH_3CCH_3
$\quad\quad\quad\overset{\|}{O}$

(ii) $CH_3CH_2CH_2CHO$

(iii) CH_3CHCHO
$\quad\quad\;\;\underset{CH_3}{|}$

68 (a) 2-hydroxypropanoic acid **[1]**

(b) CH_3CCOOH **[1]**
$\quad\;\;\overset{\|}{O}$

(c) (i) $HOOCCOOH$ **[1]**

(ii) $HOOCCH_2COOH$ **[1]**
$HOOCCH_2CH_2COOH$ **[1]**

69

	empirical	molecular	structural
a) alkanes	CH_3 C_3H_8	C_2H_6 C_3H_8	$CH_3—CH_2—CH_3$
b) alcohols	C_2H_6O C_3H_8O	C_2H_5OH C_3H_7OH	$CH_3—CH_2—CH_2—OH$ $CH_3—CH—CH_3$ 　　　　$|$ 　　　　OH
c) amines	C_2H_7N C_3H_9N	$C_2H_5NH_2$ $C_3H_7NH_2$	$CH_3CH_2CH_2NH_2$ $CH_3 CHCH_3$ 　　$|$ 　　NH_2
d) chloroalkanes	C_2H_5Cl C_3H_7Cl	C_2H_5Cl C_3H_7Cl	$CH_3—CH_2—CH_2—Cl$ $CH_3—CH—CH_3$ 　　　　$|$ 　　　　Cl
e) ketones	C_2H_6O	C_3H_6CO	$CH_3—C—CH_3$ 　　　$||$ 　　　O
f) aldehydes	C_2H_4O C_3H_6O	CH_3CHO C_2H_5CHO	CH_3CH_2CHO
g) carboxylic acids	CH_2O $C_3H_6O_2$	CH_3CO_2H $C_2H_5CO_2H$	CH_3CH_2COOH
h) amides	C_2H_5ON C_3H_7ON	CH_3CONH_2 $C_2H_5CONH_2$	$CH_3CH_2CONH_2$

70 (a) A decolorises, 1,2-dibromoethane,
CH_2BrCH_2Br **[3]**
B decolorises, 2-bromoethanol,
CH_2BrCH_2OH **[3]**
C decolorises, ethane-1,2-diol,
CH_2OHCH_2OH **[3]**

(b) from A – 1,2-dibromobutane **[1]**
from B – 2-bromobutan-1-ol **[1]**
from B – 1-bromobutan-2-ol **[1]**
from C – butane-1,2-diol **[1]**

71 (a) $HCOOCH_2CH_3$ **[1]**

(b) Show attack by lone pair from alcohol O, **[1]**
indicate that carboxylic acid carbon is more
positive than ethanol's carbon and so more
vulnerable to nucleophilic attack. **[1]**
Refer to stabilization of intermediate. **[1]**
Use arrows to show loss of H^+ from alcohol
and OH^- from carboxylic group. **[1]**

72 (a) (i)

(ii) Light is rotated in opposite directions.

(b) (i) CH_3CCOOH
　　　　$||$
　　　　O

(ii) Either optical isomer may be formed by
the addition of hydrogen across the C=O.

73 (a) chloroethane, ethanoyl chloride,
chlorobenzene **[3]**

(b) (i) chloroethane does not react **[1]**
ethanoyl chloride forms ethanoic acid.
(The reaction can be violent.)
CH_3COOH **[1]**
chlorobenzene does not react **[1]**

(ii) With NaOH solution, chloroethane forms
either ethanol **[1]** or ethene **[1]** depending
on the conditions, the hotter and more
concentrated the sodium hydroxide the
more the elimination reaction is favoured
rather than the substitution reaction. **[1]**
With cold aqueous sodium hydroxide
ethanoyl chloride will react violently, **[1]**
often after a short period when it does
not do anything. As such it is a
dangerous reaction to do. It forms
sodium ethanoate. **[1]**
Chlorobenzene will only react with
molten sodium hydroxide in a sealed
tube. **[1]** It will then form sodium
phenoxide or phenol. **[1]**

(c) The carbon in the COCl function is quite positive **[1]** as both the oxygen **[1]** and the chlorine **[1]** are more electronegative and draw electrons away from the carbon. **[1]** It is therefore susceptible to nucleophilic attack by OH^-. **[1]**
The carbon in chloroethane has one chlorine attracting electrons from it **[1]** so it has some enhanced positive nature **[1]** making it susceptible to nucleophilic attack by OH^-. **[1]**
In chlorobenzene the ring of delocalized **[1]** electrons **[1]** makes the carbon to which the chlorine is attached not susceptible to nucleophilic attack. **[1]** Only the most stringent conditions can force a nucleophile to attack the ring. **[1]**

74 (a) (i) Molar mass of aspirin $= 180$ g mol^{-1}
No. of moles of aspirin in 25 cm^3 of water $= 0.000833$.
Therefore concentration of aspirin $= 0.0333$ mol dm^{-3},
therefore $K_a = 3.0 \times 10^{-4}$ mol dm^{-3}
$$= \frac{[H^+(aq)]^2}{[0.0333]}$$
leading to $[H^+(aq)] = 3.16 \times 10^{-3}$ mol dm^{-3} **[4]**

(ii) pH $= 2.5$ **[1]**

(b) $CH_3CO_2C_6H_4CO_2H(aq) \rightleftharpoons$
$CH_3CO_2C_6H_4CO_2^-(aq) + H^+(aq)$
Increased concentration of acid in the stomach will cause this equilibrium to move to the left thus decreasing the extent of dissociation of the aspirin into ions. **[2]**

(c) As the pH increases at the stomach wall, the equilibrium shown in **(b)** moves back towards the right greatly increasing the extent of dissociation of the aspirin molecules and thus increasing $[H^+(aq)]$ to a level where they attack the stomach wall and cause bleeding. **[2]**

(d) Produce a soluble salt of 2-ethanoylbenzenecarboxylic acid (aspirin) **[1]** e.g. the sodium salt or the calcium salt (the latter is sold as 'Dispirin') by reacting it with an alkali. **[1]**

75 (a) Because citric acid has three ionizable hydrogen atoms, one on each of the $-CO_2H$ groups. The first K_a value refers to the ionization of the first hydrogen, the second to the ionization of two hydrogen atoms and the third value to all three becoming ionized. **[3]**

(b) 2.1, 4.8, 5.2 **[2]**

(c) $H_2OCCH_2C(OH)CO_2HCH_2CO_2H(aq) +$
$3NaOH(aq) \rightarrow$
$^+Na^-O_2CCH_2C(OH)CO_2^-Na^+CH_2CO_2^-Na^+$
$(aq) + 3H_2O(l)$ **[2]**

(d) Since the ionization of citric acid is an endothermic process, it will be favoured by raising the temperature. The concentration of $H^+(aq)$ ions is lower when the citric acid is refrigerated, therefore it will be less acid and so will taste less sour. **[3]**

76 (a) (i) $CaSiO_3$ **[2]**

(ii) Ca^{2+} SiO_3^{2-} **[2]**

(b) An acid/base reaction has already occurred between CaO and SiO_2 **[1]** and the solid is a salt, rather than an oxide **[1]** so it has neither acidic nor basic properties. **[1]**

77 (a) Reaction with acid involves oxidation of Mn to Mn^{2+} **[1]** which has a d^5 electronic configuration **[1]** which is particularly stable (making the reaction more favourable). **[1]**

(b) Up to Mn, the energy required to involve the 3d and 4s electrons is recouped in the compounds formed. **[1]** The nuclear charge increases as the period is crossed. **[1]** At iron the increased nuclear charge has made it impossible to recoup the energy required to involve all the 3d and 4s electrons. **[1]**

(c) Potassium is $[Ar]$ $4s^1$, copper is $[Ar]$ $3d^{10}$ $4s^1$. **[1]** The nuclear charge has increased from potassium to copper **[1]** and the shielding of the 4s electron by the 3d electrons is poor. **[1]** This makes the first ionization energy of copper (745 kJ mol^{-1}) much greater than that of potassium (418 kJ mol^{-1}), **[1]** making copper unreactive towards water because ionization of the atom requires too much energy to be recouped in the formation of an aqueous ($+1$ or $+2$) ion. **[1]**

78 (a) (i) CuF_6^{3-} **[1]**

(ii) octahedral **[1]** this shape keeps ligands as far apart as possible **[1]**

(b) (i) $+3$ **[1]**

(ii) $[Ar]$ $3d^8$ **[1]**

(iii) Two from: forms coloured complexes/ions, would be a powerful oxidizing agent, its ion would be paramagnetic. **[2]**

79 (a) Ti^{2+} and Ni^{2+} **[2]**, **(b)** Mn^{2+} **[1]**, **(c)** Cu^+ **[1]**, **(d)** Sc^{3+} **[1]**.

80 (a) (i) A and B
  (ii) D and E
  (iii) E and F
  (iv) B and C **[4]**
 (b) two isomers of C **[1]**
 (c) G and I or two isomers of D **[1]**
 (d) G and H **[1]**

81 (a) Shows alternate double **[1]** and single bonds. **[1]**
 (b) 6 p orbitals orthogonal (at right angles) to the ring. **[1]**
  π bonding system shown. **[1]**
 (c) Extended delocalization shown from ring to nitrogen, or curly arrows show lone pair feeding into ring. **[1]**
 (d) Curly arrows shown forming a double bond from the nitrogen to the ring carbon and negative charge shown on the 2 position, **[1]** the 4 position, **[1]** and the 6 position. **[1]**

82 (a) (i)
  CN
  |
  CH_3CCH_3
  |
  OH

 (ii) CH_3CH_2CHCN
  |
  OH

 (iii) [ring]CHCN
  |
  OH

 (iv) [ring]CCN
  |
  OH
  CH_3 **[4]**

 (b) (i)
  CH_2NH_2
  |
  CH_3CCH_3
  |
  OH
  1-amino-2-methylpropan-2-ol

 (ii) $CH_3CH_2CCH_2NH_2$
  |
  OH 1-aminobutan-2-ol

 (iii) [ring]CHCH_2NH_2
  |
  OH 1-amino-2-phenylethan-2-ol

 (iv) [ring]$CHCH_2NH_2$
  | CH_3
  OH
  1-amino-2-phenyl-2-methylethan-2-ol **[8]**

 (c) (i) CH_3CHCH_3 propan-2-ol
  |
  OH

 (ii) $CH_3CH_2CH_2OH$ propan-1-ol
 (iii) [ring]CH_2OH phenylmethanol
 (iv) [ring]CHCH_3   1-phenylethanol **[8]**
  |
  OH

83 A $CH_3CCH_2CH_2CO_2H$
  ‖
  O

 B $CH_3CCH_2CH_2CO_2CH_3$
  ‖
  O

 C $CH_3CHCH_2CH_2CH_2OH$
  |
  OH

 D $CH_3CCH_2CH_2CO_2H$
  ‖
  N
  | NO_2
  NH[ring]NO_2 **[4]**

84 A $CH_3CH_2CO_2H$, **B** CH_3CH_2COCl,
  X $Na_2Cr_2O_7/H_3O^+$, and distil out the product as it forms, **Y** $NH_3(aq)$ **[4]**

85 A CH_3CHCH_3
  |
  OH

 B CH_3CCH_3
  ‖
  O

 C OH
  |
  CH_3CCH_3
  |
  CN

 D OH
  |
  CH_3CCH_3
  |
  CO_2H

 E [CH_3
  |
  $C - CH_2$
  |
  CO_2CH_3]$_n$

 F CH_3CH_2CHO

 G $CH_3CH_2CH{=}NNH$[ring]$^{NO_2}_{NO_2}$
 H CH_3CH_2CHCN
  |
  OH

 I $CH_3CH{=}CHCO_2C_2H_5$
 J [CH_3
  |
  $CH - CH$
  |
  $CO_2C_2H_5$]$_n$

 K CH_3
  \
   $C{=}NNH$[ring]$^{NO_2}_{NO_2}$
  /
  CH_3

 X P_2O_5
 Y H_3O^+, 370 K **[13]**

86 (a) m/e 136 is the parent ion **[1]** the peak at m/e 91 ($C_7H_7^+$) suggests a phenyl ring with an alkyl side chain **[1]** the IR suggests an OH group **[1]**, a phenyl group **[1]**, a $C{=}O$ group **[1]** and a $C{-}O$, perhaps an ester. **[1]** The NMR data are complex, but along with the information above can be assigned to phenylethanoic acid **[1]** as follows:

NMR spectrum	11.8δ (singlet, 1 proton), C$_6$H$_5$—CH$_2$—CO**OH** **[1]**
/ppm	7.1δ (singlet, 5 protons), **C$_6$H$_5$**—CH$_2$—COOH **[1]**
	3.6δ (singlet, 2 protons), C$_6$H$_5$—C**H$_2$**—COOH **[1]**

(b) (i) C$_6$H$_5$—CH$_2$—COOCH$_3$ **[1]** methyl phenylethanoate **[1]**

(ii) C$_6$H$_5$—CH$_2$—COCl **[1]** phenylethanoyl chloride **[1]**

(iii) C$_6$H$_5$—CH$_2$—CH$_2$OH **[1]** phenylethanoic acid **[1]**

(c) the C=O peak in the IR would decrease in intensity/disappear. **[1]**

Appendix 1

Table of elements

Name	Symbol	Atomic Number	Approximate molar mass/g mol^{-1}	Molar mass /g mol^{-1} for the naturally-occurring isotopic composition to maximum precision so far obtained
Actinium	Ac	89	227	227.0278
Aluminium	Al	13	27	26.98154
Americium	Am	95	243	(243)
Antimony	Sb	51	122	121.75
Argon	Ar	18	40	39.948
Arsenic	As	33	75	74.9216
Astatine	At	85	210	(210)
Barium	Ba	56	137	137.33
Berkelium	Bk	97	247	(247)
Beryllium	Be	4	9	9.01218
Bismuth	Bi	83	209	208.9804
Boron	B	5	11	10.81
Bromine	Br	35	80	79.904
Cadmium	Cd	48	112.5	112.41
Caesium	Cs	55	133	132.9054
Calcium	Ca	20	40	40.08
Californium	Cf	98	251	(251)
Carbon	C	6	12	12.011
Cerium	Ce	58	140	140.12
Chlorine	Cl	17	35.5	35.453
Chromium	Cr	24	52	51.996
Cobalt	Co	27	59	58.9332
Copper	Cu	29	63.5	63.546
Curium	Cm	96	247	(247)
Dysprosium	Dy	66	162.5	162.50
Einsteinium	Es	99	254	(254)
Erbium	Er	68	167	167.26
Europium	Eu	63	152	151.96
Fermium	Fm	100	257	(257)
Fluorine	F	9	19	18.998403
Francium	Fr	87	223	(223)

Table of elements

Name	Symbol	Atomic Number	Approximate molar mass/g mol^{-1}	Molar mass /g mol^{-1} for the naturally-occurring isotopic composition to maximum precision so far obtained
Gadolinium	Gd	64	157	157.25
Gallium	Ga	31	70	69.72
Germanium	Ge	32	72.5	72.59
Gold	Au	79	197	196.9665
Hafnium	Hf	72	178.5	178.49
Helium	He	2	4	4.00260
Holmium	Ho	67	165	164.9304
Hydrogen	H	1	1	1.0079
Indium	In	49	115	114.82
Iodine	I	53	127	126.9045
Iridium	Ir	77	192	192.22
Iron	Fe	26	56	55.847
Krypton	Kr	36	84	83.80
Lanthanum	La	57	139	138.9055
Lawrencium	Lw	103	260	(260)
Lead	Pb	82	207	207.2
Lithium	Li	3	7	6.941
Lutetium	Lu	71	175	174.97
Magnesium	Mg	12	24	24.305
Manganese	Mn	25	55	54.9380
Mendelevium	Md	101	258	(258)
Mercury	Hg	80	200.5	200.59
Molybdenum	Mo	42	96	95.94
Neodymium	Nd	60	144	144.24
Neon	Ne	10	20	20.179
Neptunium	Np	93	237	237.0482
Nickel.	Ni	28	59	58.70
Niobium	Nb	41	93	92.9064
Nitrogen	N	7	14	14.0067
Nobelium	No	102	259	(259)
Osmium	Os	76	190	190.2
Oxygen	O	8	16	15.9994
Palladium	Pd	46	106.5	106.4
Phosphorus	P	15	31	30.97376
Platinum	Pt	78	195	195.09
Plutonium	Pu	94	244	(244)

Table of elements

Name	Symbol	Atomic Number	Approximate molar mass/g mol^{-1}	Molar mass /g mol^{-1} for the naturally-occurring isotopic composition to maximum precision so far obtained
Polonium	Po	84	209	(209)
Potassium	K	19	39	39.0983
Praseodymium	Pr	59	141	140.9077
Protactinium	Pm	91	231	231.0359
Radium	Ra	88	226	226.0254
Radon	Rn	86	222	(222)
Rhenium	Re	75	186	186.207
Rhodium	Rh	45	103	102.9055
Rubidium	Rb	37	85.5	85.4678
Ruthenium	Ru	44	101	101.07
Samarium	Sm	62	150.5	150.4
Scandium	Sc	21	45	44.9559
Selenium	Se	34	79	78.96
Silicon	Si	14	28	28.0855
Silver	Ag	47	108	107.868
Sodium	Na	11	23	22.98977
Strontium	Sr	38	88	87.62
Sulphur	S	16	32	32.06
Tantalum	Ta	73	181	180.9479
Technetium	Tc	43	97	(97)
Tellurium	Te	52	127.5	127.60
Terbium	Tb	65	159	158.9254
Thallium	Tl	81	204	204.37
Thorium	Th	90	232	232.0381
Thulium	Tm	69	169	168.9342
Tin	Sn	50	117	118.69
Titanium	Ti	22	48	47.90
Tungsten	W	74	184	183.85
Uranium	U	92	238	238.09
Vanadium	V	23	51	50.9414
Xenon	Xe	54	131	131.30
Ytterbium	Yb	70	173	173.04
Yttrium	Y	39	89	88.9059
Zinc	Zn	30	65	65.38
Zirconium	Zr	40	91	91.22

Appendix 2

Mass spectrum data

Peaks due to charged ions formed as a result of fragmentation		
Peak observed in the mass spectrum (m/e)	Possible identity of the ion responsible	Plausible inference about the parent ion (m)
29	CHO^+	–
30	$CH_2NH_2^+$	Primary amine
31	CH_3O^+	–
41	$C_3H_5^+$	–
43	$C_3H_7^+$	–
43	CH_3CO^+	Methyl ketone
44	CO_2^+	–
44	$C_3H_8^+$	–
45	$COOH^+$	Carboxylic acid
57	$C_2H_5CO^+$	Ethyl ketone
59	$CH_3CO_2^+$	Methyl ester
73	$C_2H_5COO^+$	Ethyl ester
77	$C_6H_5^+$	C_6H_5-R (R = alkyl)
91	$C_7H_7^+$	C_6H_5-R (R = alkyl)

Peaks resulting from the loss of uncharged groups from the parent ion			
Peak observed in the mass spectrum (m/e)	Group lost from molecule giving rise to the parent ion (M)	Likely identity of molecule giving rise to the parent ion (M)	
M−16	NH_2	Amide	$R-CONH_2$
M−17	OH	Alcohol	R-OH
M−18	H_2O	Alcohol	R-OH
M−28	CO	Ketone	R-CO-R′
M−29	CHO	Aldehyde	RCHO
M−29	C_2H_5	Ethyl ketone	C_2H_5CO-R
M−30	NO	Nitroarene	$C_6H_4R-NO_2$
M−31	OCH_3	Methyl ester	$R-CO-OCH_3$
M−32	CH_3OH	Methyl ester	$R-CO-OCH_3$
M−41	C_3H_5	Propyl ester	$R-CO-OC_3H_7$
M−42	C_3H_6	Butyl ketone	C_4H_9-CO-R
M−43	CH_3CO	Methyl ketone	CH_3-CO-R
M−44	CO_2	Carboxylic acid Ester	R-COOH R-CO-OR′
M−45	COOH	Carboxylic acid	R-COOH
M−46	C_2H_5OH	Ethyl ester	$R-CO-OC_2H_5$
M−46	NO_2	Nitroarene	$C_6H_4R-NO_2$
M−59	CO_2CH_3	Methyl ester	$R-CO-OCH_3$

Appendix 3

Infra-red data

Wave number range	Group	Vibration
3750–3200	Alcohols	O—H stretching
3500–3300	Amines	N—H stretching
3500–3140	Amides	N—H stretching
3300–2500	Carboxylic acids	O—H stretching
3095–3010	Alkenes	C—H stretching
3030	Arenes	C—H stretching
2962–2853	Alkanes	C—H stretching
2260–2215	Nitriles	C≡N stretching
1795	Acid chlorides	C=O stretching
1750–1735	Esters	C=O stretching
1740–1720	Aldehydes	C=O stretching
1725–1700	Carboxylic acids	C=O stretching
1700–1680	Ketones	C=O stretching
1700–1630	Amides	C=O stretching
1669–1645	Alkenes	C=C stretching
1650–1590	Amides and amines	N—H bending
1600, 1500 and 1450	Arenes	C=C stretching
1485–1365	Alkanes	C—H bending
1310–1100	Esters and alcohols	C—O stretching
880–700	Arenes	C—H bending
800–600	Chloroalkanes	O—Cl stretching

Appendix 4

NMR data
Typical proton chemical shift values (δ)

Type of proton	Chemical shift (ppm)
R—CH$_3$	0.9
R—CH$_2$—R	1.3
R$_3$CH	2.0
CH$_3$—C(=O)—OR	2.0
R—C(CH$_3$)(=O)	2.3
C$_6$H$_5$—CH$_3$	2.3
R—C≡C—H	2.6
R—CH$_2$—Hal	3.2–3.7
R—O—CH$_3$	3.8
R—O—H	4.5*
RHC=CH$_2$	4.9
RHC=CH$_2$	5.9
C$_6$H$_5$—OH	7*
C$_6$H$_5$—H	7.3
R—C(=O)—H	9.7*
R—C(=O)—O—H	11.5*

*Sensitive to solvent, substituents, concentration.
All values relative to T.M.S. = 0